T3-BOE-114

RELIGION AND SEXUAL HEALTH

Theology and Medicine

VOLUME 1

RELIGION AND SEXUAL HEALTH

Ethical, Theological, and Clinical Perspectives

Edited by

RONALD M. GREEN

Department of Religion, Dartmouth College,
Hanover, New Hampshire, U.S.A.

KLUWER ACADEMIC PUBLISHERS
DORDRECHT / BOSTON / LONDON

Library of Congress Cataloging-in-Publication Data

Religion and sexual health : ethical, theological, and clinical perspectives / edited by Ronald M. Green.
p. cm. -- (Theology and medicine ; v. 1)
Includes bibliographical references and index.
ISBN 0-7923-1752-1 (hard : alk. paper)
1. Sex--Religious aspects--Christianity. 2. Sexual ethics.
3. Sex (Psychology) 4. Sex therapy. I. Green, Ronald Michael.
II. Series.
BL65.S4R45 1992
291.5'66--dc20 92-10060

ISBN 0-7923-1752-1

Published by Kluwer Academic Publishers,
P.O. Box 17, 3300 AA Dordrecht, The Netherlands.

Kluwer Academic Publishers incorporates
the publishing programmes of
D. Reidel, Martinus Nijhoff, Dr W. Junk and MTP Press.

Sold and distributed in the U.S.A. and Canada
by Kluwer Academic Publishers,
101 Philip Drive, Norwell, MA 02061, U.S.A.

In all other countries, sold and distributed
by Kluwer Academic Publishers Group,
P.O. Box 322, 3300 AH Dordrecht, The Netherlands.

Printed on acid-free paper

Printed in the Netherlands

TABLE OF CONTENTS

RONALD M. GREEN

INTRODUCTION

For many people today sexuality remains an area of deep inner conflict. Modern developments in medicine, biology, and psychology have transformed – even revolutionized – our sexual self-understanding. In society, the struggle for liberation by women and gay persons has challenged longstanding and deeply ingrained ideas about gender and sexuality. Yet, beneath the surface of many persons' sexual self-awareness lie attitudes and norms shaped by centuries of religious teaching. Although these attitudes and norms frequently run counter to more contemporary information this does not always lessen their force or their influence in people's lives.

This anomalous situation evokes an image offered nearly a century ago by the philosopher Nietzsche. In his *Joyful Wisdom*, Nietzsche describes a madman who, descending from the mountains to a village square, announces, "God is dead. God remains dead. And we have killed him." The madman is puzzled when his repeated cries draw no response from the townfolk. Suddenly, he understands:

> I have come too early. I am not yet at the right time. This prodigious event is still on its way, and is still travelling, – it has not yet reached men's ears. Lightning and thunder need time; the light of the stars needs time; deeds need time, even after they are done, to be seen and heard. This deed is as yet further from them than the most distant star, – *and yet they have done it!* ([1], pp. 168f.).

In terms of sexuality, we are much like Nietzsche's townfolk. For over a century, science and social experience have shed new light on our understanding of sexuality, sexual normalcy and sexual health. As a culture, we have vastly distanced ourselves from many assumptions about sexuality that our forebears took for granted. Yet for many – if not most – people the light has not yet arrived. In our innermost being we continue to function with attitudes toward our body and our sexuality shaped in childhood by parents, teachers, and peers who themselves were shaped by ideas rooted in the religious past. Not all these ideas are wrong. Some represent enduring wisdom about sexuality and personal life found in the Bible and other scriptures of our religious traditions. But many religious ideas about sexuality derive from intellectual environments no more relevant to our own than Hebraic or Ptolemaic cosmology is to the world of modern astronomy.

When people try to live their sexual lives by conforming to outdated or erroneous religious values or norms, the result is often the kinds of sexual dysfunction and maladjustment seen by sex counselors and other therapists. Sometimes, the specific religious factors in such dysfunction are obvious.

Ronald M. Green (ed.), Religion and Sexual Health, vii–xiv.

Sometimes, they are harder to detect since they arise at the deepest levels of patients' existence and are hidden even to them. Further complicating things is the fact that religiously-induced inhibitions and fears can contribute to sexual problems not only in obviously religious people or active churchgoers but also in persons who regard themselves as secular or religiously "emancipated." This hidden quality of religious influences on sexuality can hamper counselors or therapists seeking to help patients. Counselors also may be unaware of the diverse ways religious beliefs impinge on sexuality and they may be professionally unequipped to address the specific religious beliefs and attitudes that influence sexual functioning and sexual health.

This volume aims to bring to the surface and clarify the complex relationship between religion and healthy sexual adjustment. It does so in several ways. First, it seeks to shed light on the heritage of religious sexual teachings that continue, often in subtle and less than obvious ways, to influence our contemporary sexual attitudes and behaviors. Second, it seeks to suggest the complexity of this heritage and to accentuate its positive as well as its negative implications. Sexually maladjusted persons are often influenced by partial or erroneous interpretations of the sexual teachings of their religious traditions. Counselors, by understanding the full complexity of these traditions, can orient themselves and their patients toward alternative religious approaches that can assist sexual adjustment.

Finally, this book seeks to offer a series of diverse clinical perspectives, including case studies, involving conflicts between religion and sexual health. These perspectives are meant to assist counselors and clinicians deal with religiously influenced sexual problems. They illustrate the complex ways in which religious beliefs enter into sexual maladjustment and can even render patients averse to therapy. These clinical views are also meant to be a resource for ethicists, theologians, and philosophers working with the conceptual heritage of religious traditions and struggling to reorient these traditions to modern realities.

The first division of the volume, "Religion and Sexual Ethics" begins with a historical essay, "Christianity and Sexuality," in which Vern L. Bullough offers a comprehensive overview of the development of Christian sexual teachings. Bullough does not mince words. Almost from its start, he tells us, Christianity was "a sex negative" religion that barely tolerated even the minimal expression of sexuality for procreative purposes. Whatever residue of acceptance of the body and sexuality the early Christian theologians absorbed from their Hebraic and New Testament roots was quickly washed away by Stoic and Neoplatonic 'idealism' that regarded sexuality as one of the "lower appetites" in need of constant discipline and self-control. This idealism was accentuated by Christianity's encounter with forms of Gnostic and Manichaean dualism, whose myths led to the view of sexual reproduction as a perpetuation of humankind's primal fall from spiritual perfection. In the writings of St. Augustine, this volatile intellectual tradition culminated in what was to become the normative Christian view of sexuality: sex was (just barely) permissible within marriage for the sake of procreation while all intentionally non-procreative forms of sex were regarded as sinful. As Bullough's essay makes clear, by the end of the fifth century of the common era these complex intellectual struggles led to a normative tradition that was averse to sex and the body, that held celibacy as the ideal, that had no place for non-procreative sexual expression, and that was hostile to homosexuality.

Augustine's position formed the core of most Christian teachings on sexuality until the beginning of our century. Although Protestant theologians and ethicists have increasingly distanced themselves from many aspects of this teaching, Roman Catholicism still officially adheres to major aspects of the traditional view. Speaking from within the Catholic tradition, theologian Charles Curran identifies five continuing problem areas for the tradition: its propensity to regard sexual matters in terms of negative dualisms; its patriarchal approach to sexuality and gender; its focus on legal considerations; its authoritarianism; and its reliance on a non-historical, "classicist" and "physicalist" conception of natural law. The Catholic tradition, in Curran's view, is not without its positive features, as well. Offsetting these negative elements, Curran identifies five aspects of the tradition that can be used to update the tradition and form a Catholic contribution to enhanced sexual adjustment for individuals today. These include the tradition's positive assessment of all features of the human, including sexuality; its deep methodological confidence that in religious-ethical matters faith and reason cannot contradict one another; its recognition that scripture can and should be supplemented by experience, especially ongoing historical and cultural realities; its communitarian approach that runs counter to a bare individualism and perceives people as existing in networks of relationships; and its potential for alternate models of the teaching office of the Church that would replace hierarchy and authoritarianism with more democratic and participatory forms of ethical leadership. Curran concludes with a point of great importance for the counseling and therapeutic setting: the observation that, whatever its specific teachings, Catholic moral thinking has always given individual conscience primacy in matters of ethical conduct and decision.

In his essay "Body Theology and Human Sexuality," James Nelson continues themes suggested in the two previous essays. He offers a listing of what he calls the "seven deadly sins" that have marked Christian teaching about sexuality. This list includes spiritualistic dualism; patriarchal dualism; heterosexism; guilt over self-love; legalistic sexual ethics; a sexless image of spirituality; and the privatization of sexuality. Nelson seeks to overcome this unfortunate heritage in two ways. First, he brings to the fore seven "virtues" or resources within the tradition that counter this heritage of anti-sexualism and oppression. Second, he calls for the development of a new kind of spiritual enterprise he terms "body theology." Instead of assuming that religious traditions exist merely to "teach us" how properly to conduct ourselves sexually, body theology takes our physical being and sexuality as important sources of ethical-religious insight. Nelson concedes that because our bodily experience is itself partly the complex expression of cultural and social forces, it will contain many negative elements from our religious heritage, and he points to male bodily experience as a particular area of malaise and alienation. Nevertheless, he maintains that a theology wishing to contribute to sexual health for individuals and society must begin by moving deeply into our body experiences.

Curran's critique of hierarchical and authoritarian forms of sexual teaching in Roman Catholicism is deepened by Barbara Andolsen in her essay "Whose Sexuality? Whose Tradition? Women, Experience, and Roman Catholic Sexual Ethics." Andolsen calls for a new approach to the tradition of Catholic and Christian sexual teaching, one that pays special attention to the historical experiences of women and less ecclesiastically powerful men. Heeding these often silenced voices, Andolsen contends, not only calls into question the deeply misogynistic themes of authoritative teaching, but may also throw new, critical light on some modern sexual-moral ideals like the cult of romantic love. Andolsen concludes that if contemporary Catholic and other religious ethicists are to provide resources for healing our sexual "brokenness"and for facilitating sexual wholeness, they face several tasks: they must work out the implications of the shift away from procreation as the telos that justifies sexual activity; they must identify violent or coercive sexual activity and unequal distributions of social power as fundamental sexual evils; and they must replace older emphases on the restraint of sexual evil with a focus on sexual satisfaction and the enhancement of sexual intimacy.

George Edwards, a New Testament scholar, provides an extended illustration of the uses – and misuses – of scripture in relation to one of the most pressing sexual moral issues of our day: the ecclesial and synagogal inclusion of uncloseted lesbians and gay men. While acknowledging the culture-boundedness of scripture and its interpreters, Edwards nonetheless argues that neither the Hebrew Bible nor the New Testament provides a basis for the kinds of contempt with which Jews and Christians have so often treated homosexual persons. Edwards believes that the Sodom story of Genesis 19 is badly misinterpreted if it is taken as a condemnation of homosexuality or if it is read backwards into scripture and used to reinforce the claim that Genesis 1–3 establishes heterosexuality as normative. Paul's argument in Romans 1 is also misread if it is taken as a criticism of gentile homosexual practices. Rather, Edwards maintains, it forms part of Paul's larger argument for a stance of "faith righteousness" that renders invalid all forms of self righteousness based on some human achievement or qualification. According to Edwards, a well grounded understanding of the New Testament sources replaces the "we/them" distinctions so often employed to justify discrimination against homosexuals with a powerful message of "foundational inclusivity" for homosexual persons.

Christine Gudorf's essay calls attention to the western religious traditions' support of the patriarchal family structure, an aspect of religious teaching that is often neglected but that has profound and pervasive implications for the whole realm of sexual functioning. Although the New Testament offers some countervailing motifs, Gudorf observes that patriarchy has been the dominant model in

Jewish and Christian culture and has contributed to the existence of three standard features of the Western family: the headship/breadwinner role of husbands/fathers; wives' subordination to husbands and restriction to motherhood; and the imperative that children obey parents. Each of these features has a bearing on people's emotional health and (eventual) sexual adjustment. For example, the emphasis on male headship contributes to men's serious problems of depression, anxiety, and emotional isolation, and is an important factor in the epidemic of abuse suffered by women and children in families. Other features of this family structure contribute to dependency needs in women and children and lifelong problems in the establishment of intimacy between spouses. Overcoming the many emotional and sexual problems generated by the family, Gudorf suggests, will require us to confront the internal contradictions concerning the family found at the heart of our religious traditions.

The essay by William Stackhouse that forms the second division of this book provides some empirical confirmation for the theological critiques that precede. Based on the results of a 1985 internal program development study conducted by the United Church of Christ, it reports the sexuality experiences of over 2,800 UCC congregants. The study asked respondents to address thirty-four areas of sexuality experience and to comment on the adequacy of religious and other support services in helping them address these areas of concern. Narrative responses gathered in the study revealed six problem areas: adolescent and young adult sexuality-related experiences; experiences concerning homosexuality; experiences with extramarital relationships; marital problems related to sexual dysfunction and desire; problem pregnancy experiences; and experiences with sexual abuse, rape, and incest. In facing these problems, respondents commonly reported feelings of isolation and "abnormality" made even more acute by a lack of information, support, or counsel from religious and other sources. Some respondents who faced marital difficulties reported feeling that there was "no place to go," while others who sought help found that the counselors they saw were not adequately prepared or responsive. A consistent finding of the study was that respondents desired counsel from clergy but also wanted a higher level of expertise and sensitivity in their counselors.

The third and clinical division of the book begins with an essay by Julian Slowinski, whose theological training has equipped him to perceive the religious roots of forms of sexual maladjustment he encounters in his work as therapist and counselor. Against the background of a review of the history of religious teaching in this area, Slowinski observes that recent and more enlightened developments in Christian sexual teachings have not filtered down to many patients. These people often continue to work with a "grammar school understanding" of their faith tradition's teachings on sexuality. Sometimes, this baggage of misconceptions is a source of serious psychological harm. This is particularly true for many HIV positive and AIDS patients, whose suffering is compounded by religiously-induced guilt. Religious misconceptions also play a role in the cases of individuals or couples whose ability to experience sexual pleasure is inhibited by guilt or religiously based mistrust of self-gratification.

Slowinski argues that for patients experiencing sexual problems in which religious belief or training is a factor, there can be value in a "holistic" approach to therapy that draws on modern religious ideas, presents biblical teaching in context, or places sexuality within an incarnational framework.

It is natural to assume that sex counselors and other therapists may tend to develop a negative view of the impact of religious teaching on sexuality partly because they only see religious persons who are experiencing sexual problems. Although this association must be kept in mind, the essay by William Simpson and Joanne Ramberg reveals a dynamic working in the opposite direction and one that may actually cause us to underestimate the adverse impact religious fears and inhibitions can have on sexual adjustment. As Simpson and Ramberg point out, religious concerns may lead sexually troubled people to avoid professional assistance in the first place or to flee therapy prematurely. In the course of a discussion of six cases in which religious factors play a role in the formation of sexual problems, Simpson and Ramberg observe this pattern of religiously-induced evasion or flight. Among other things, their essay aims at sensitizing counselors to this problem and alerting them to the deep and sometimes non-obvious ways that religious backgrounds can hinder the therapeutic process.

Harold Lief's essay, "Sexual Transgressions of Clergy," deals with a problem that lies at the vortex of religion and sexual maladjustment. As Lief's discussion makes clear, clergy are not immune to sexual problems and transgressions. It may even be that the unrealistic expectations imposed on them, along with the special powers and opportunities for intimacy the clerical role affords, render ministers, priests or rabbis especially vulnerable to sexual wrongdoing. Lief's essay documents these dynamics in a series of cases involving Episcopal priests. Issues in these cases range from unwelcome experiences of disclosure of closeted gay priests through other clergymens' involvement in forms of sexual abuse or abuse of the pastoral role. As Lief makes clear at the close of his discussion, the sexual transgressions of clergy raise important questions for religious traditions as they confront human sexuality. Which elements of religious teaching – especially in the area of gender roles – contribute to the kinds of abuse he discusses? How do outdated sexual ideas and teachings, whether in seminary curricula or elsewhere, contribute to pastors' misconduct or parishioners' vulnerability to it? And what can be done at the doctrinal, educational, and disciplinary levels to avoid these problems?

The essay by Bishop David Richards presents the opposite and more positive side of the pastoral encounter with sexual problems. It develops the ways in which changing visions of sexuality and the ministry have affected the role of the pastoral counselor. Richards observes that where pastoral counseling once focused largely on providing short-term, situational support and comfort for people in crisis, it now increasingly involves dealing with the need for intellectual and spiritual growth on the part of individuals who form the pastor's constituencies. Corresponding to this change is a movement from a focus on sexual rules or violations of norms – a position that lives on in the "just say 'no'" movement with regard to sexual ethics – toward more constructive ways of integrating sexuality with the whole of life. Richards observes that this more demanding task requires a new standard of clinical and theological education for clergy in sexual matters, both to help them understand their sexuality and to enable them to assist congregants in the task of sexual self-development.

William R. Stayton begins his essay,"Conflicts in Crisis" with an observation that is central to this book. According to Stayton, we have learned more in the last thirty years about human sexual behavior and human sexual response than we have ever known, but almost all this knowledge conflicts with traditional religious belief systems. Stayton goes on to identify and discuss five major areas of sexual conflict resulting from these changes. They include sexual identity, sex-coded roles, sexual orientation, sexual lifestyle, and sexual response and function. In each of these areas, Stayton believes, the degree of conflict between religion and sexual knowledge will depend on which of two religious approaches to sexual ethics or sexual theologies one adopts. One is an Act-centered theology emphasizing specific right or wrong forms of conduct; the other is a Relationship-centered approach focusing on motives and consequences. In Stayton's view it is not possible, as many people would like to do, to combine these approaches since they inherently contradict one another. Clinical experience shows that people pursue myriad possibilities for sexual and emotional gratification. Because of this, both patients and counselors must become clear about the approach they adopt. Stayton contends that for a therapist, a Relationship-centered approach is best suited to helping people deal with the complexity of their sexual lives.

Sister Margretta Dwyer closes the volume with a brief reflection, partly based on her work in treatment programs with sex offenders. As an illustration of the way in which some rigidly interpreted traditional religious beliefs can undermine sexual health and personal or social sexual adjustment, she mentions the case of a man in an unhappy marriage who practices incest with his daughters because he accepts his religion's teaching that extramarital relationships are wrong. Dwyer goes on to place thinking of this sort in a mythological context shaped by religion and medicine. The latter, she points out, has for centuries developed mistaken notions of human sexuality that religions have picked up and sanctified. Long after science has corrected these erroneous views, they continue to operate at the core of our beings, leading to fears and repressions that sometimes break out in forms of self-destructive or antisocial sexual behavior. Dwyer ends the volume by hoping for the private and collective revelations needed to light up our minds.

Mention of the arrival of light returns us to Nietzsche's madman, who was right to proclaim, "God is dead – And we have killed him" if we interpret this to mean that modern scientific research has invalidated major aspects of religious teaching in the area of sexuality. What is remarkable is how recent these developments really are. Although this process had clearly begun in Nietzsche's time, with the pioneering work of people like Krafft-Ebing or Sigmund Freud, as William Stayton reminds us it is not yet thirty years since Masters and

Johnson undertook the first clinical investigations of human sexual functioning. Only in the last decade have we made progress in understanding the complex role biological factors may play in sexual orientation. In many respects, sexual ethics and sexual theology must begin anew if they are properly to assimilate this information.

Yet the essays in this volume also show that Nietzsche's madman was wrong to believe that God is dead, if we mean by this that people have ceased to bring deep spiritual and religious resources to bear on their lives as a whole and their sexuality in particular. As the essays in this volume show, many people continue to interpret their sexual being in religious terms. Sometimes this has the unfortunate consequences seen in the therapeutic setting. But sometimes it provides an opportunity for sexual self-development and understanding. Whichever is the case, the essays in this volume suggest that religion cannot be bypassed by all those interested in promoting sexual health and adjustment. Whether for good or bad, whether consciously or unconsciously, sexuality is often placed in a religious context. The choice then is usually not between religion and sexual adjustment, but between outdated religious ideas and newer forms of sexual theology and ethics that contribute to sexual health.

I want to thank the contributors to this volume for their efforts. Coming from a diversity of professional backgrounds, they were remarkably able to focus their attention on our common concerns. I also want to thank the series editor, Earl Shelp, for his counsel and unfailing patience. Earl's own work spans so many of the concerns of this volume that he was a major resource in my thinking about it. My own years of teaching in the field of religion and sexuality convince me that successfully joining these crucial areas of life remains one of humankind's major challenges.

Dartmouth College
Hanover, New Hampshire, U.S.A.

BIBLIOGRAPHY

1. Nietzsche, F.: 1910, *The Joyful Wisdom*, T. Common (trans.), Vol. 10 of *The Complete Works of Nietzsche*, T.N. Foulis, Edinburgh.

SECTION I

RELIGION AND SEXUAL ETHICS

VERN L. BULLOUGH

CHRISTIANITY AND SEXUALITY

Traditional western Christianity was a sex negative religion, regarding sex as necessary for procreation, but emphasizing celibacy as the ideal. This hostility to sex is not so much a part of Biblical Christianity as it is a major component of Christian theology. This was because Christianity appeared in a world dominated by the expansionist Roman Empire which had adopted and incorporated into its own intellectual tradition much of Greek philosophy and ethics.

In fact one of the reasons for the ultimate success of Christianity was that it managed to incorporate much of this intellectual tradition into its own theology, particularly those intellectual trends which were so dominant in the early period of Christian development. Even the conflicts in this post-Hellenistic classical thinking were carried over into Christianity and it was these conflicts which were a major factor in disagreements among Christians and which gave rise to conflicting interpretations and what might be called sectarianism. Ultimately emerging as dominant in the west was a combination of Stoic and Neoplatonic ideals which were given their ultimate Christian version by St. Augustine. Essentially this view held that the ideal life was one of celibacy but that God had also recognized marriage as long as it was entered into for the purpose of procreation. Sexual intercourse was permissible only if procreation was the intent, the correct orifice and instrument were used (vagina and penis), and intercourse was in the proper position (with the male on top and the female on the bottom). This historical review examines how these intellectual traditions which were so influential in the first few centuries of the modern era were incorporated into western Christianity.

THE BIBLE AND SEX

Scriptural references to sex in both the Jewish and Christian scripture are not particularly numerous. Adultery was punishable by death if the couples were caught in the act (Exodus 20:14; Deuteronomy 5:18; Leviticus 20:10 and elsewhere). Though this was modified by Jesus in his rescue of the woman being stoned for adultery (John 7:53), other parts of the Christian scripture also accept the death penalty (e.g. Mark 10:19; Romans 13:9). Adultery was also given a new definition in the Christian scriptures. A divorced person who remarried while the first spouse was still living committed adultery (Mark 10:10–12; Luke 16:18; Matthew 19:3–9).

Sexual relations with an animal were condemned and the perpetrator was to be put to death, as was the animal (Exodus 22:19; Leviticus 18:23; 20:15–16;

Ronald M. Green (ed.), Religion and Sexual Health, 3–16.

Deuteronomy 27:21). Homosexuality has often been interpreted as being condemned in the story recounting the destruction of Sodom and Gomorrah (Genesis 19) although this interpretation has been challenged by numerous modern commentators. Moreover, a somewhat similar account about the town of Gibeah (Judges 19) did not result in its immediate destruction. The issue is not, however, so much what modern commentators of the Bible regard as the real meaning of this particular biblical passage, but the fact that it was read throughout much of the history of western Christianity as a condemnation of homosexuality, with the term Sodomy becoming equated with homosexual activity in both English and American legal codes. Moreover, those interpreting the sodom story as hostile to homosexuality could also rely upon the legal opposition to male homosexuality as expressed in Leviticus, which calls for the death penalty (18:22; 20:13). Traditionally St. Paul, too, was read as being opposed to homosexuality. In 1 Corinthians 6:9 Paul indicates that effeminate men, idolators. and adulterers will not inherit God's kingdom, while in Romans 1:27, he condemns men "committing shameless acts with men," as well as women who "exchange natural relations for unnatural" to the "inevitable penalty" for their perversion, although again alternative meanings have been given to the relevant passages by modern commentators as is discussed later in this volume.

Transvestism is condemned (Deuteronomy 22:5), as is masturbation, although this interpretation is again subject to dispute by modern commentators. The source of the condemnation is the Genesis story where Er, the son of Judah, dies before he can impregnate his wife Tamar. This leaves it up to Onan, Er's brother, to fulfill the Levirate requirement and impregnate his brother's wife. Onan, however, ejaculates on the ground rather than putting his seed into her. It is unclear whether he masturbated or practiced *coitus interruptus* (Genesis 38). Traditional interpretations indicate that he masturbated. The fact that most of the Christian Church fathers ignored Onan's defiance of an expressed command of the Hebrew God indicated just how much they tried to read sanctions against sexuality into every ambiguous scriptural passage.

This Christian hostility to sex is emphasized by the fact that Judaism itself, in spite of such scriptural verses, did not regard sex as essentially evil but instead perceived sex as a pleasurable activity, particularly within the confines of marriage. Women as well as men were free to enjoy sex and it was part of a husband's marital obligation to pleasure his wife (Deuteronomy 24:5).

This brief review is not the place to develop more fully Jewish or Biblical attitudes toward sex. In general, these might be summed up by stating that although certain forms of sexual activity were prohibited, sexual intercourse, at least within marriage, was to be enjoyed ([7], pp. 74–92). Christian attitudes, however, go far beyond the Biblical sources and are based upon classical philosophy, particularly Stoicism and Neoplatonism, as well as some of the religions that competed with Christianity during the first three centuries of its existence.

STOICISM

The Stoics considered sex a special type of pleasure and felt that men (and women) made poor use of their time if they occupied their minds with such matters. Though sexual enjoyment in itself was morally indifferent, sex, like wealth, was not a worthy goal for reasonable adults to seek, and the pursuit of sexual pleasure was not conducive to a healthy morality (Seneca, *On Benefits* [*De beneficiis*] 7.2.2 and *Epistolae* 104:34; also ([1], p. 348), and ([9], p. 54)). Sex, however, was not bad in itself, since reproduction involved the soul as well as the material body. The resultant seed was in part a divine thing, a fragment of a sort of soul plasma torn from the spirits of our ancestors. Reason was not, however, present in the embryo but developed only after birth. In this sense human beings were animals, but what ultimately distinguished them from animals whose impulses were governed only by desire and aversion, was their ability to govern themselves by reason. Sexual intercourse belonged in the category of the "lower appetites," in which the wise man refrained from indulging. The truly wise person cultivated a sober and reserved demeanor, and indulged in any of lower concerns of the body like eating, drinking and sexual activity only to the minimum essential for bodily health (Epictetus, Discourses, 2.1815–18, 3:22.76 and *Encheiridion,* 33:8).

The Stoic watchwords were nature, virtue, decorum, and freedom from excess. Immoderation in bodily activities was irrational because it made a person dependent on his or her own body. Marriage was recognized, but passion in marriage was suspect, because the only justification for marriage was the propagation of the race. The first century A.D. Stoic teacher Mussonius Rufus went so far as to teach that marital intercourse was permissible only if the purpose was procreative. Intercourse for pleasure within the confines of marriage was reprehensible. Since homosexual activities were for pleasure alone, not for reproduction, they too were condemned and classed as unnatural ([20], pp. 71–77). Seneca, the first century A.D. Stoic rhetorician and statesman, was cited by St. Jerome as claiming that a

> wise man ought to love his wife with judgment, not affection. Let him control his impulses and not be borne headlong into copulation. Nothing is [more] foul than to love a wife like an adulteress. Certainly those who say that they united themselves to wives to produce children for the sake of the state and the human race ought, at any rate, to imitate the beasts, and when their wife's belly swells not destroy the offspring. Let them show themselves to their wives not as lovers, but as husbands (*Fragments,* No. 85).[1]

NEO-PLATONISM

Also influential on Christianity were the Neoplatonists who adopted somewhat similar ideas about sexuality. Particularly influential in this respect was the Alexandrian Jewish philosopher, Philo, born in the last quarter of the first century B.C. As a Jew, Philo accepted the Jewish belief in the divine command to procreate and replenish the earth. From this belief it followed that marriage was blameless and worthy of high praise, but Philo, heavily influenced by the ascetic philosophy of his time, held that sex in marriage could only be justified if the goal of the husband and the wife was the procreation of legitimate children for the perpetuation of the race. Philo described as mere pleasure lovers those who mated with their wives, not to beget children, but like "pigs or goats" in quest of sexual enjoyment (On the *Special Laws* [*De Specialibus Legibus*], III, 113). His hostility to sex without progeny led him to state that those men who mated with barren women were worthy of reproach, for in their seeking after mere pleasure, "they destroyed the procreative germs with deliberate purpose" (*On the Special Laws,* III, 34–36). Predictably all kinds of non-procreative sex were condemned by those adopting Philo's reasoning, particularly homosexuality. Philo regarded the male homosexual as enslaved to irrational passion and infected with the female disease:

> In former days the very mention of it was a great disgrace, but now it is a matter of boasting not only to the active but to the passive partners, who habituate themselves to endure the disease of effemination, let both body and soul run to waste, and leave no ember of their male sex-nature to smoulder .... These persons are rightly judged worthy of death by those who obey the law, which ordains that the man-woman who debases the sterling coin of nature should perish unavenged ... And the lover of such may be assured that he is the subject of the same penalty (*On the Special Laws,* III, vii, 37–42).[2]

The most influential of the Neoplatonists was Plotinus, who lived and worked in the third century of the modern era. After studying in Alexandria, he moved to Rome, where he had great influence on the Emperor Gallienus and his wife, Salonina. Plotinus was a religious mystic and Plotinism (or Neoplatonism) is a theocentric form of thought. Porphyry, the pupil and biographer of Plotinus, recounted how four times, while in a state of ecstasy, Plotinus was made one with God. Plotinus is also said to have been so ashamed of his body that he considered his parentage and birthplace of no importance. His mysticism and piety were, however, dependent on reason and intellectual balance. So determined was Plotinus to retain his independence that he refused to affiliate himself with any organized worship on the grounds that the gods must come to him, not vice versa. He differed from Plato in his greater stress on religious and mystical orientation, going so far as to insist that the nature of the Real was attainable only in a state of mystical ecstasy from which the last trace of not only sensible but also intelligible experience had been erased. He believed that there was no personal immortality; rather, the goal of human life was to merge with the universal spirit. The path of redemption was long and gradual, taking aeons of reincarnation to traverse and requiring long and careful training (*The Enneads,* v. 3, par 1–9; I, 6, par 9). Though the body and its needs were not to be despised, they had to be disciplined in such a way that nothing distracted the soul from the contemplation of higher things. The core of human virtue was in detachment from worldly goods. This indifference was essential to put an individual out of reach of the caresses and stings of material life. By implication it was important to become indifferent to sex. This indifference is especially noticeable in the writings of his pupil Porphyry. In his work *Abstinence from Animal Food* Porphyry condemned any kind of pleasure as sinful, including horseracing, theatergoing, dancing, eating meat, and, of course, sexual intercourse under any conditions (I, 45; IV, 8, 20).

CHRISTIAN RIVALS

The ascetic and philosophical concepts of Stoicism and Neoplatonism as well as other classical philosophical concepts influenced not only Christianity but also its rivals, particularly Gnosticism and Manichaeanism. Because Christianity was competing from early in its development with Gnosticism and other redemptionist cults, and later with Manichaeanism, it was greatly influenced by what its rivals said or taught. In fact, within any particular Christian community, what Christian spokesmen said about sex was very heavily influenced by what its rivals said on the same topic. The fact that these competing groups had many of the same ideas about sexuality as were developed by Christianity is indicative, I think, of how persuasive such ideas were. Some went to greater ascetic extremes than Christianity, and it might well be that Christianity succeeded because it was perceived by many as somewhat more moderate, but this "moderation" was relative and dependent on any particular Christian community. Some Christian communities were particularly noted for their ascetic practices. The physician Galen (second century A.D.), for example, observed that the Christian community in Rome included men and women who, like the philosophers, refrained from "cohabitating all their lives" ([21], p. 65).

Gnosticism, an early major rival and in many ways a variant version of Christianity, also had many different strands, some of which had greater influence on western Christianity than others. Key to Gnostic belief was the conception of dualistic worlds, one evil and material, the other good and spiritual. Humans had elements of both the good and the evil; their purpose on earth was to seek redemption which came through a saving *gnosis* or knowledge secretly revealed to their predecessors and transmitted to others by the initiate alone. This knowledge concerned the supreme God, superior to the Creator, known only to those who, as spiritual beings, had originally emanated from this one. Recognition of the One and of themselves would save them so that after death they could escape this alien world of their Creator to join with the supreme God. In the meantime, their spirits had been temporarily imprisoned in fleshly bodies. The key to salvation was to free the body from its bondage.

Such a doctrine often led to a stringent asceticism, for the best way a true Gnostic could express his or her alienation from simply human existence was by adopting an ascetic life, and most particularly by abstaining from sex. In a sense, Gnosticism was, as a modern commentator has stated, a mixture of "Christian theology and sexual morality" ([15], p. 58). Not all gnostics were ascetics, however. Some demonstrated their indifference to the pleasures of life by rejecting sexual asceticism. Two different and contradictory arguments were put forth to justify such conduct. Some said that human actions could never be subject to moral law since they represented the moral or earthly side of the individual and not the soul; others argued that actions usually considered sinful were not sinful for true believers. Most of what we know about the Gnostics comes from Christian writers who were trying to point out their errors. One group of Gnostics followed Marcion, a one time Christian who had been excommunicated by the Christian community in Rome in 144 A.D. and as a result set up a separate competing organization. Marcion taught that nature was evil because it was created out of evil matter, and since his followers did not wish to fill the world with other evil matter, they [the Marcionites] abstained from sexual intercourse and from marriage (Clement, *Stromata,* III, cap 3 (12)).[3] According to Marcion, Jesus came as a life-giving spirit to manifest a new revelation as well as new way of life but his message had been distorted by false apostles under the spell of Judaism. Only Paul and Luke had understood the true gospel but errors had even crept into their teaching, errors that Marcion thought he was able to eliminate. The Marcionites taught that sex was evil, as was reproduction and growth. Jesus, they said, had descended from heaven as a fully formed adult without undergoing birth, boyhood, or temptation. All Marcionite believers were to remain celibate (Tertullian, *On the Flesh,* cap I; *Against Marcion,* IV, cap vii, and V, cap vii). Sharing the ascetic outlook of Marcion was Julius Casianus who taught that Jesus had been sent to earth to stop individuals from copulating (Clement, *Stromata,* III, cap 17 (102)).

Some of the more orthodox Christians such as Justin Martyr became heavily influenced by the Gnostic version of Christianity. Though a convert to Christianity, Justin described approvingly a Christian youth who asked the surgeons to emasculate him as a protection for bodily piety and pointed with pride to those Christians who renounced marriage to live in perfect continence (*Apology,* I, cap 29). Justin was so committed to the idea that sex was evil that he could not believe that Mary had sexually conceived Jesus. He further argued that Mary had been undefiled and conceived as a virgin, and he made her the antitype of Eve with whom he associated sexual intercourse (*Dialogue with Trypho,* 100). Even those Christians who married for the sake of begetting children were advised by Clement of Alexandria to control their will and be chaste when conception was not possible (*Stromata,* II, vii). Perhaps one of the reasons that gnostic ideas penetrated so deeply into Christianity is that most of the Church Fathers were unmarried, and even those who, like Tertullian, were married, tended to denigrate marriage. Tertullian came to feel such deep remorse over his lapse into matrimony that he joined the Montanists, a heretical Christian sect that emphasized celibacy (*To His Wife,* I, ii-iii). Tatian, a disciple of Justin Martyr who converted to Gnosticism after the martyrdom of his teacher, taught that sexual intercourse had been invented by the Devil, and thus anyone attempting to be married was trying to do the impossible: serve two masters, God and the Devil (Clement, *Stromata,* III, cap 12 (8)). At the other extreme were the followers of Nicolas, an early Christian who drew condemnation from the author of the Book of Revelations (2:6, 14–15) for his antinomianism. His followers taught that women were to be regarded as common property and that believers could have intercourse as they wanted and with whom they wished (Clement, *Stromata,* III, cap 2 (5–9), cap 4 (25–26). Other Gnostics were accused of regarding sexual intercourse as a sacred religious mystery, the knowledge and practice of which would bring them to the kingdom of God. Such teachings were based on a scriptural passage not now found in the Bible:

All thing were one, but as it seemed good to its unity not to be alone, an inspiration came from it, it had intercourse with it, and it made the beloved (Clement, *Stromata,* III, cap 2 (5–9), cap 4 (29)).

In general, the Gnostics seem to have emphasized asceticism rather than antinomianism. Some of the accusations against them might simply have been Christian rhetoric but it matters little to us today whether they taught such things or not; Christians became victims of the rhetoric and it had a great deal of influence upon Christian attitudes. By the end of the second century the Gnostics were declining. The organizational ability of the more orthodox Christians had begun to win the battle against them. Christians accomplished this by emphasizing the importance of the community as opposed to the individual and by emphasizing scriptural antecedents while at the same time incorporating pagan philosophy. In the process, however, the Christian Church became thoroughly imbued with the Neoplatonic and Stoic ideals as well as the teachings of some of its rivals; its sexual attitudes, for example, were much more influenced by the ascetic Gnostics than by the more earthly Jews or even particular Christian scriptural references.

The general temper of the times, whether Gnostic or Christian, favored continence. Soranus of Ephesus, probably the most renowned medical writer of the second century of the modern era, believed that permanent virginity was the most healthful state of life, since

Even among animals we see that those females are stronger which are prevented from having intercourse. And among women we see that those who, on account of regulations and service to the gods, have renounced intercourse and those who have been kept in virginity as ordained by law are less susceptible to disease. If, on the other hand, they have menstrual difficulties and become fat and ill-proportioned, this comes about because of idleness and inactivity of their bodies .... Consequently permanent virginity is health, in male and female alike; nevertheless, intercourse seems consistent with the general principle of nature according to which both sexes (for the sake) of continuity (have to ensure) the succession of living beings (*Gynecology,* I, vii (32)).

Sexual asceticism continued to have a strong hold over the Christian mind. Gregory of Nyssa, in the fourth century, in his *On Virginity* dismissed marriage as a sad tragedy. St. Jerome (d. 420 A.D.) emphasized the inconvenience and tribulations of married life (*Against Helvidius,* 21–22; also *Letters,* xxii)[4] and summarized his views in an oft-quoted passage:

I praise marriage and wedlock, but I do so because they produce virgins for me. I gather roses from thorns, gold from the earth, and pearl from the shell ([11], XXII, 20).

St. Ambrose (d. 397 A.D.) called marriage a "galling burden" (*On Widows [De Vidius*], cap XIII, xxxi; and *On Virginity,* I, cap 6) and urged all those contemplating matrimony to think about the bondage and servitude into which wedded love degenerated (*On Widows,* cap XV, lxxxviii; and cap XI, lxix). Derek Sherwin Bailey summed it up by stating that the Fathers argued with monotonous regularity that the wedded state was not as good as the single, and though most could not bring themselves to say that marriage was an evil, they were perfectly willing to count it as only thirtyfold compared with the sixtyfold of widowhood and the hundredfold of virginity ([3], p.24).

There was Christian opposition to this cult of virginity, but that the works of the three most vocal defenders of marriage – Jovinian, Helvidius, and Vigilantius – have not survived provides some indication of the hostility of the Christian establishment to their position, and their arguments are known only through their opponents, particularly St. Jerome, one of the formulators of the western Christian belief system. Jerome went to great lengths to refute them, although he was careful to point out that he did not condemn wedlock itself. He did admit, however, that he had great difficulty in understanding why people would want to be married (*Against Jovinian, Against Vigilantius*; also *Against Helvidius,* 21–22). With such cavalier dismissals of the advocates of marriage, celibacy carried the day as the ideal state for Christians. It became a common practice among many Christian groups to forbid marriage after ordination to the priesthood, although there was at first general agreement that a man might be married before ordination. Inevitably there was a suggestion that matrimonial cohabitation disqualified a person entirely for priestly ministration. The Council of Nicæa specifically rejected an absolute rule of clerical celibacy (Sozomen, *Ecclesiastical History,* I, xxiii; Socrates, *Ecclesiastical History,* I, xi), but later Councils, particularly that of Trullo in the seventh century, turned to the subject again. By the fifth century, in fact, the Western Christian Church tended to hold that bishops, presbyters, deacons, and others employed before the altar were to refrain from coitus, although mitigating circumstances occasionally might allow for exception ([3], p. 30). The Eastern Christian Church never adopted such a position.

The growing difference between the Western and Eastern positions on celibacy and sex are probably due more to the influence of St. Augustine (d. 430) than to any other factor. The hostility to sexuality so omnipresent in the Western Christian Church might have gradually weakened with the decline of Gnosticism if it had not once again been threatened by the appearance of a new rival, Manichaeanism, which again emphasized that sexual abstinence was a key to salvation. Manichaeanism was based on the teachings of the prophet Mani (216–277) who lived and was crucified in southern Babylonia. His religion incorporated various aspects of Gnosticism, Stoicism, Christianity, Neoplatonism, and Zoroastrianism. Before his death, his teachings had become influential in Egypt, Palestine, and Rome, and from there spread throughout the Roman Empire. Manichaeanism had a canonical scripture (the seven books of Mani), claimed to be a universal religion, and had a hierarchy and apostles. In short, it had all the attributes of Christianity. Like Muhammad after him, Mani taught that he was the last prophet in a chain of revelations. Manichaeanism was a missionary faith combining science, philosophy, and religion into a new dualistic synthesis. Although claiming the authority of revelation, the Manichaeans also paid great deference to reason. The universe was divided into two pantheistic portions, the kingdoms of Light and of Darkness, in juxtaposition to each other, with each reaching out into infinity. Light and Darkness were both eternal and uncreated powers in everlasting opposition and conflict, although the God of Light was alone able to know the future. Eventually, Light would overcome Darkness, but the ultimate victory depended not on the defeat of Darkness, but on the withdrawal of Light.

Originally the two realms had existed without intermingling, but history began when the Prince of Darkness, attracted by the splendor of the Light, invaded the domain of Light. In a cosmic battle the forces of Light had been vanquished and then devoured by their opponents, just as had happened in Greek mythology to Dionysus and the Titans. The result was that Light was imprisoned. The God of Light then sent forth His Word, or the Living Spirit, into the universe to create the earth, the moon, the sun, the planets, and the 12 elements or creative eons.

Adam and Eve were the result of a union between a son and daughter of the Prince of Darkness who still retained some light. Adam had more light than Eve. Recognizing their miserable fate, Adam and Eve begged for help from the God of Light who thereupon sent Jesus, the Incarnate Word, to warn Adam that Eve was the tool of Darkness and urge Adam to reject her blandishments. Adam successfully did so but Eve copulated with an earth Archon (a son of the God of darkness) and gave birth to Cain, Abel and two daughters, resulting in two married pairs. Abel's wife soon became pregnant, but Abel, knowing that he had not made her pregnant, accused his brother of having sexual intercourse with her and in the subsequent argument, he was slain by Cain. Eve, grieving over the loss of her son, was taught witchcraft by the earth Archon so that she was able to seduce Adam and give birth to Seth, who was filled with such great elements of light that the Archon conspired to destroy him. Adam managed to foil them by taking Seth away and, in due time, the God of Light sent His Word, the Christ, this time to accomplish the redemption of the Elect.

This rather complex and far fetched mythology of creation proved attractive to many of the time. Sexually, it was important to the development of Christianity because procreation was regarded by the Manichaeans as an evil act that perpetuated the cycle of imprisoning light. The purpose of man was to gain light, and this could be done by eating bread, vegetables, or fruit containing seeds. The growth of light could also be affected or impeded by sexual actions, since the seed of man contained light. Those entering the Manichaean religion were supposed to have tamed concupiscence as well as covetousness and not only remained celibate, but refused to eat flesh or drink wine. Christianity rejected the dualism and the specific mythology, but the sexual asceticism so much a part of Manichaeanism was retained primarily through the works of St. Augustine.

Manichaeans divided the world into three classes. The first were the Adepts – the Elect – who believed and followed the teachings of Mani by renouncing private property, practicing celibacy, observing strict vegetarianism, and never engaging in trade. The second were the Auditors, men and women of good will who could not yet fully contain their material desires and who earned money, owned property, ate flesh and even married. Still they could ultimately earn high rewards by serving and supporting the Elect in this life and always striving to become an Elect. Finally, there were the completely sensual members of society, totally lost in wickedness, who rejected the gospel of Mani. At death, the Elect went directly to the Paradise of Light, the Auditors spend a period of purification in purgatory, and the wicked were doomed to eternal and irrevocable suffering in the three Manichaean Hells.[5]

St. Augustine was an adherent of the Manichaean cause for some eleven years, but never reached the Elect stage, in part because of his difficulties with sex. He remained an Auditor, living with a mistress and growing more and more uncomfortable about his inability to control his lustful desires. His own ambivalence about the matter is evident from his constant prayer: "Give me chastity, and continency, but do not give it yet" (*Confessions,* VIII, vii). Augustine's mother, Saint Monica, though tolerating her son's concubine, and even his son by her, Adeodatus, wanted Augustine to become a Christian and to enter into marriage.

Ultimately and reluctantly Augustine concluded that the only way his venereal desires could be satisfied was through marriage (*Confessions,* VI, xii; also *The Happy Life* [*De beata vita*], I (4)). He sent his mistress and son away and became engaged to a girl who was not yet of age (twelve was the legal minimum, fourteen was the custom) but while waiting to be married, Augustine found that he still could not refrain from sex and took another mistress. At this juncture he went through a personal crisis that ended only with his conversion to Christianity. Once he became a Christian, he found himself also able to adopt a life of celibacy. Triggering the crisis was his reading of a passage in the New Testament:

> Not in rioting and drunkenness, not in chambering and wantonness, not in strife and envying, but put ye on the Lord Jesus Christ, and make not provision for the flesh, to fulfill the lusts thereof (Romans, 13:13–14.).

Augustine interpreted this as a call to celibacy. Conversion for him came to mean the rejection of sexual intercourse.

> So thou convertedst me unto thyself, as that I sought now no more after a wife, nor any other hopes in the world (Augustine, *Confessions,* VIII, xii.).

Having accepted continence as the most desirable model for life, Augustine became particularly offended by the act of coitus. He wrote that he knew nothing that brought "the manly mind down from the heights [more] than a woman's caresses and that joining of bodies without which one cannot have a wife" (*On the Nature of Good,* [*De natura boni*] cap xvii). He was particularly upset that generation could not be accomplished without what he felt was a certain amount of "bestial" movement (*On the Grace of Christ, and on Original Sin, against Pelagius,* [*De gratia Christi et de peccato originali*], cap 43, xxxviii) and violent lustful desires (*City of God,* XIV, 26). In short, he adopted most of the sexual teachings of the Manichaeans. Sexual lust, he argued, was an inevitable result of the expulsion of Adam and Eve from the Garden of Eden since before their expulsion they had been able to control their genitals, which had been obedient to the dictates of their will and never stirred except at their behest (*City of God,* XIV, 26). Adam and Eve, in fact, had not engaged in sexual intercourse before their expulsion, although if they had chosen to do so, they could have managed the affair without lascivious heat or unseemly passion (*City of God,* XIV, 23, 24, 26; *On Marriage and Concupisence* [*De nuptiis et concupiscentia*], II, cap 14 (9)).

Once Adam and Eve had fallen from Paradise, they became conscious of the new impulse generated by their act of rebellion, and this drove them to an insatiable quest for self satisfaction ([3], p. 54). Augustine termed this impulse *concupiscence* or *lust,* and it was through this that the genitals lost the docility of innocence and were no longer amenable to the will. In this, he was also following Manichaean ideas since Mani had taught that sexual sin consisted not only in the overt act of sex itself, but also in the impulse. Inevitably, according to Augustine, Adam and Eve had felt shame over their desires, a shame causing them to cover their nakedness by sewing fig leaves together to make aprons concealing their genitals. Concupiscence was, however, still displayed through the sexual impulses, which proved stronger and less tractable than other passions and could be satisfied only through an orgasm engulfing the rational faculties in violent sexual excitement. Though coitus must be regarded as good, since it came from God, every concrete act of intercourse was evil, with the result that every child could literally be seen to have been conceived in the sin of its parents (*On the Merits and Remission of Sins* [*De peccatorium meritis et remissione*], cap 57 (XXIX)). Venereal desire, as implanted by God for encouraging the increase of mankind, must be regarded as blameless, but the same desire corrupted by concupiscence was to be regarded as shameful and sinful. Because generation could not occur unless the carnal union of husband and wife was motivated by the stimulus of lust, it must, therefore, be a sin (*On Marriage and Concupisence,* I, cap 4 (iii)).

Did this make marriage an evil in itself? Augustine tried to distinguish between matrimony and sexual intercourse, but his answers were rather contradictory. He concluded that concupiscence could not take away the good of marriage and that marriage mitigated somewhat the evil of concupiscence. He summed up his argument by stating that we "ought not to condemn marriage because of the evil of lust, nor must we praise lust because of the good of the marriage" (*On Marriage and Concupisence,* I, cap 8 (vii)). Marriage, however, did manage to transform coitus from a satisfaction of lust to a necessary duty (*On Marriage and Concupisence,* I, cap 9 (viii)). This was so because the sexual act, when employed for human generation, lost some of its inherent sinfulness (*On the Good of Marriage* [*De bono conjugali*], X, 10; Sermons, CCCLI, iii (5)), though it still remained the channel by which the guilt of concupisence was transmitted from parents to children. The guilt could be removed only by a baptismal regeneration, even though the impulse of lust and sense of sexual shame still remained (*Against Two Letters of the Pelagians [Contra duas epistolas pelagianorum*], I, cap 27 (xiii), cap 30 (xv)). Nonetheless, marriage guarded wanton marital indulgence from the grave sinfulness of fornication or adultery and even though nuptial embraces were not always intentionally destined for procreation, the sin resulting from this was only venial providing there was no attempt to frustrate the natural consequences of coitus [i.e., pregnancy] (*On Marriage and Concupisence,* I, 17 (xv)). Sexual intercourse should, however, only take place with the instrument and orifice designed for that purpose by God, with the penis and vagina, and in the proper, male superior position. All intercourse between the unmarried was condemned by Augustine, although he held that true wedlock could exist without ceremony. In short, the only justification for intercourse was procreation and only between husband and wife (*On the Good of Marriage,* V, and *Soliloquies,* I, 10 (17)). Celibacy was the highest good, and intercourse in itself was only an animal lust, but in marriage, and only in marriage, it was justified because of the need for procreation. Even then it could be evil unless procreation ultimately came about and the proper positions and orifices were used.

Augustine's influence was the dominant one on the western church. Church officials strove mightily to inculcate the Augustinian ideals into their members. It was outlined in the penitential literature.[6] Ultimately, however, these ideals were incorporated into Canon law which essentially institutionalized the Augustinian view [6].

Though most Christians undoubtedly engaged in sexual activities, as evidenced by the fact that the population in medieval Europe increased over the generations, they did so knowing that unless procreation resulted, they were probably engaging in a sinful act. This guilt feeling was complicated by the widespread belief that a woman also contributed seeds to making the fetus and was only able to do so when she reached orgasm. This view was advanced by Galen whose medical ideas were dominant in the medieval and early modern period. Galen taught that the woman was colder than the male and to engage successfully in sexual activities had to be warmed up. Galen said:

The mixing of the two sperms is necessary for two useful purposes. The first is that the woman's sperm is a suitable source of sustenance for the sperm of the man, because the sperm is thick and of a hot constitution, whereas the sperm of the woman is thin and of a cold constitution. Because of its thickness, the sperm of the man cannot spread sufficiently and through its heat it would spoil the substance of which the foetus is made, the sperm of the woman is thus necessary to moderate its thickness and heat. Its second use is the formation of the second membrane surrounding the foetus. For the man's sperm, moving forward in a straight line, does not reach the horn-like extensions and does not spread out over the whole internal surface of the womb. So the sperm of the woman is necessary to reach the places where the sperm of the man has not reached ([8], p. 397).

Though some medieval writers were sceptical about the necessity of the female orgasm since they had interviewed women who claimed to have become pregnant without having an orgasm, the assumption was that it was necessary to bring the woman to orgasm in order to follow the Augustinian injunction of utilizing sex only to beget offspring.

In sum, the Christian Church brought an overlay of sinfulness to almost every aspect of human sexuality. Masturbation, fornication outside marriage, homosexuality, transvestism, adultery, and in fact almost any aspect of sexual behavior was sinful and ultimately against church law. How successful was the Christian Church in enforcing their attitudes and laws? Probably they were not particularly successful over all, but what they did was breed a deep feeling of guilt about sexual activity which remains one of the more troubling aspects of the Christian heritage. Though there were modifications of the basic teachings by various Protestant writers, and a general weakening of religious influence in the nineteenth and twentieth centuries, the guilt feelings remain, although more in some versions of Christianity than in others. These guilt feelings did not result from any particular teaching of Jesus but were in fact adopted into Christianity by the Doctors of the western Church who were reflecting the zeitgeist of their own time and in the process imbued in millions of people today a guilt feeling that does not necessarily correspond to modern understanding of sexuality.

SUNY College
Buffalo, New York, U.S.A.

NOTES

[1] The passage is found in St. Jerome, *Against Jovinian*, (1:30). Haase ([19], p. 10) claims the passage came from a lost treatise by Seneca on marriage, *De matrimonio*.
[2] See also ([2], p. 58).
[3] An English translation of the relevant passage can be found in [16].
[4] See also ([11], LIV, 4) and ([12], p. xlvii).
[5] For a discussion see [18]. The writings of St. Augustine are the best extant source for the teachings of Mani, particularly his *The Way of Life of the Manichaeans* [*De Moribus Manichaeorum*]. Also valuable is the *Fihrist, or Register of the Sciences of Muhammad ibn Ishaq ibn al-Nadim* in the tenth century in Arabic. For an English translation, see [10].
[6] See [17], ([7], pp. 355–387); An English translation of many of the penitentials can be found in [14]. The Irish penitentials have been edited by Beiler [4]. There is also a collection in German [22]. On the relationship of the penitentials to German folk law see [5].
[7] For further discussion of this see ([13], esp. pp. 201–204).

BIBLIOGRAPHY

1. Arnold, E.V: 1911, *Roman Stoicism,* reprint ed. 1958, Routledge and Kegan Paul, London.
2. Baer, R.A., Jr.: 1970, *Philo's Use of the Categories Male and Female,* E.J. Brill, Leiden, Netherlands.
3. Bailey, D.S.: 1959, *Sexual Relation in Christian Thought,* Harper, New York.
4. Beiler, L. (ed.): 1963, *The Irish Penitentials,* Dublin Institute for Advanced Studies, Dublin.
5. Berman, H.J.: 1983, *Law and Revolution: The Formation of Western Legal Tradition,* Harvard University Press, Cambridge, MA.
6. Brundage, J.: 1987, *Law, Sex, and Christian Society in Medieval Europe,* University of Chicago Press, Chicago.
7. Bullough, V.L.: 1976, *Sexual Variance in Society and History,* University of Chicago Press, Chicago.
8. de Koning, P. (trans.): 1903, *Trois traites d'anatomie arabes,* E.J. Brill, Leyden, Netherlands.
9. Hadas, M.: 1968, *The Stoic Philosophy of Seneca,* W.W. Norton, New York.
10. Ibn al-Nadim, M.: 1970, *The Fihrist of al-Nadim,* B. Dodge (ed. and trans.), 2 vols., Columbia University Press, New York.
11. Jerome: 1933, *Select Letters,* F.A. Wright (ed. and trans.), William Heinemann, London.
12. Jerome: 1949, *Lettres,* J. Labourt (ed. and trans.), Société les belles lettres, Paris.
13. Lemay, H.R.: 1982, 'Human Sexuality in Twelfth-through Fifteenth-Century Scientific Writings', in V.L. Bullough and J. Brundage (eds.), *Sexual Practices and the Medieval Church,* Prometheus Books, Buffalo, pp. 187–205.
14. McNeill, J.T. and Gamer, H.M. (eds. and trans.): 1938, *Medieval Handbooks of Penance,* Columbia University Press, New York.
15. Noonan, J.T., Jr.: 1966, *Contraception: A History of its Treatment by Catholic Theologians and Canonists,* Harvard University Press, Cambridge, MA.
16. Oulton, J.F.L. and Chadwick, H.: 1954, *Alexandrian Christianity,* Westminster Press, Philadelphia.
17. Payer, P.: 1984, *Sex and the Penitentials,* University of Toronto Press, Toronto.
18. Puech, H.-C.: 1949, *Le Manicheisme,* Civilisations du sud [S.AEP], Paris.
19. Seneca: 1853, *Opera,* Frederich G. Haase (ed.), Teubner, Leipzig.
20. Van Geytenbeck, A.C.: 1963, *Musonius Rufus and Greek Diatribe,* B. L. Hijamans, Jr. (trans.), Van Gorcum & Company, Assen, Netherlands.
21. Walzer, R.: 1949, *Galen on Jews and Christians,* Oxford University Press, London.
22. Wasserschleben, F.W.H.: 1851, *Die Bussordnungen der abendlandischen Kirche,* reprint ed. 1958, Graz, Austria, Akademische Druck-U-Verlagsantal.

CHARLES E. CURRAN

SEXUAL ETHICS IN THE ROMAN CATHOLIC TRADITION

The Roman Catholic tradition in sexual ethics and sexual understanding has had a long history and a great influence on people and attitudes both within and outside the Roman Catholic Church down to the present day. However, at the present time the Catholic tradition and teaching are being questioned by many Catholics themselves.

The general outlines of the official Catholic teaching on sexuality are well known. Genital sexuality can be fully expressed only within the context of an indissoluble and permanent marriage of male and female, and every sexual act must be open to procreation and expressive of love union. The natural law theory supporting such an understanding results in an absolute prohibition of artificial contraception, artificial insemination even with the husband's seed, divorce, masturbation, homosexual genital relations, and all premarital and extramarital sexual relationships. Virginity and celibacy are looked upon as higher states of life than marriage. Women are not allowed to be priests or to exercise full jurisdiction within the church.

Dissatisfaction with official Catholic sexual teaching came to a boiling point when Pope Paul VI in his encyclical *Humanae Vitae* in 1968 reiterated the condemnation of artificial contraception for Catholic spouses. Before late 1963, no Catholic theologian ever publicly disagreed with Catholic teaching condemning artificial contraception. But events in the church (especially Vatican Council II, 1963–65) and the world very quickly created a climate in which many Catholic married couples and theologians publicly called for change in the official teaching condemning artificial contraception. After much consultation and hesitation, Pope Paul VI in 1968 reiterated the condemnation. The encyclical occasioned widespread public theological dissent from the papal teaching [18].

Many Catholic couples disagreed with the teaching in practice. According to the statistics of the National Opinion Research Center in 1963, 45 percent of American Catholics approved artificial contraception for married couples; whereas in 1974, 83 percent of American Catholics approved such contraception ([14], p. 35). Archbishop John Quinn of San Francisco at the 1980 Synod of Bishops in Rome gave the statistics that 76.5 percent of American Catholic married women of childbearing age use some form of contraception, and 94 percent of these women were employing means condemned by the pope [31]. Andrew Greeley concluded that the issuance of *Humanae Vitae* "seems to have been the occasion for massive apostasy and for a notable decline in religious devotion and belief." Greeley attributes the great decline in Catholic practice in the United States in the decade of 1963–73 to the teaching of this encyclical ([14], pp. 152–154).

Ronald M. Green (ed.), Religion and Sexual Health, 17–35.

Dissatisfaction with official Catholic teaching on sexual meaning and morality has been raised both in theory and in practice with regard to masturbation, divorce, and homosexuality. Many Catholic women have become disenchanted with the Catholic Church because of its attitudes and practices concerning the role of women in the church. The patriarchal reality of the Catholic Church today is quite evident. Abortion has recently become a very heated topic in Catholic circles. One important aspect centers on law and public policy, but the moral issue of abortion has also been raised. Although most Catholic theologians and ethicists remain in general continuity with the Catholic traditional teaching on abortion, some have strongly objected to this existing teaching.

A widespread dissatisfaction with hierarchical Catholic sexual teaching exists within Roman Catholicism at the present time [13]. In general, I share that dissatisfaction, but my dissatisfaction with hierarchical Catholic teaching does not involve the acceptance of an impersonal, individualistic, and relativistic understanding of sexuality which is too often proposed in our society today. The purpose of this study is not to deal with all the specific issues mentioned above or with any one of them in particular or in depth. Instead, I will try to explain the negative elements in the Catholic tradition that have influenced the existing teaching and will then appeal to other aspects in the tradition that can come up with what I would judge to be a more adequate sexual ethic and teaching.

PROBLEM ASPECTS OF THE ROMAN CATHOLIC TRADITION

This section will briefly discuss five aspects of the Roman Catholic tradition in sexual ethics which in my judgment have had a negative effect on the teaching – negative dualisms in the tradition, a patriarchal understanding, an overriding legal concern, an overly authoritarian approach to the teaching office, and the supporting natural law method.

Negative Dualisms. The Catholic tradition in sexuality has suffered from negative dualisms of both a philosophical and a theological nature. Platonic philosophy which was often used in the early church looked upon matter and corporality in general and sexuality in particular as inferior to spirit and soul. Theological dualisms often associated the bodily and especially the sexual with evil and sin. The preceding essay by Vern Bullough explains in great detail the dualisms in early Christianity which downplayed the corporeal and bodily aspects of sexuality and the sexual pleasure connected with it.

The influence of this early development on subsequent Catholic tradition and teaching cannot be denied. Until the Second Vatican Council Catholic teaching proposed that procreation and education of offspring is the primary end of marriage. The secondary ends of marriage are the mutual help of the spouses and the remedy of concupiscence basically found in Augustine ([21], pp. 192–224). In his 1930 encyclical, *Casti Connubii*, Pope Pius XI also recognized the secondary end of conjugal love and gave more importance to the personalist values of marriage. The American moral theologians, John C. Ford and Gerald Kelly, writing in 1963, acknowledged that some want to speak of sexual fulfillment rather than a remedy of concupiscence, but they opted for the older term ([12], pp. 16–165, especially 97–98). A more adequate understanding came to the fore at the Second Vatican Council, but its relationship to the older approach is still debated ([21], pp. 192–224).

Patriarchal Approaches. Contemporary feminist writings have correctly pointed out the patriarchy present in Roman Catholic life and thought. The hierarchical magisterium continues to insist that it is God's will that only males be admitted to the ministry of priesthood, but in the judgment of many such a refusal shows the continuing deeply ingrained patriarchy at work in Roman Catholicism. I believe without any doubt that the primary internal church issue facing the Roman Catholic Church in the United States now and in the immediate future concerns the role of women in the church. The existence and influence of patriarchy in Roman Catholicism has been amply discussed elsewhere ([33], pp. 24–45). This section will develop some aspects of patriarchy as it affects the Catholic approach to human sexuality.

The subordination of the female to the male in sexuality as well as in all aspects of life has been a part of the Catholic tradition and has basically gone unquestioned until recently. The negative dualism in Catholic sexual understanding has been used to support the subordination of the woman to the man in marriage as in life. From earliest times the woman was identified with the bodily, the corporeal, the emotional, and the material. The male was identified with the spirit and rationality [34].

Catholic emphasis on synthesis, unity, and the hierarchical ordering necessary to bring about societal unity also influenced the subordinate role given to women. Catholic social theory has insisted that human beings are by nature social and political. For individuals to come together in a true society there is need for authority to organize and bring cohesiveness and unity to the community. Until comparatively recent times hierarchical organization and authority were thought to be necessary for society [32].

Marriage and family form a very significant social human community. Here too there is need for cohesiveness and unity which according to the Catholic tradition are brought about by a hierarchical ordering of the community of marriage and the family. Pope Pius XI in his 1930 encyclical, *Casti Connubii*, still used such a paradigm to understand marriage and the family. Different roles and functions are assigned so that a unified whole results but with the husband having the ultimate authority and power. According to Pope Pius XI the husband has authority over his wife and children while ready submissiveness and willing obedience must characterize the wife in accord with the command of the Apostle Paul. However, the submission should not infringe on the freedom and dignity of the wife or oblige her to yield to unreasonable or incompatible desires of her husband. The wife is not on the same level as children. The pope makes use of an analogy which frequently appeared in the recent Catholic tradition – the husband is the head of the domestic body while the wife is its heart. The analogy of the human body with its different parts serves to understand better the different roles that people play in marriage with final authority resting with the husband as the head ([30], pp. 126–127).

Anthropological dualism and a hierarchical ordering combined in the early church to distinguish various degrees of women in the church. Virgins, widows, and wives constituted a hierarchical ordering of the role of women. Emphasis on the superiority of virginity is still found in contemporary documents of the hierarchical church ([29], pp. 29–30). In the early church virginity took over the preeminent role that had been given earlier to martyrdom. Until recently Catholic liturgical documents referred to women saints as martyrs, virgins, or neither martyrs nor virgins. Thus there was no positive explanation for those who were neither martyrs nor virgins.

Patriarchy and its effect on sexual understanding were immeasurably strengthened by the fact that men had the authority and power in the society and were the canonists and theologians who wrote about these matters. The celibate male perspective naturally colored the whole approach to human sexuality.

From my perspective there is one caveat that should be entered here. I do not think that people who have never experienced something should be excluded from thinking and writing about it. We who have not experienced hunger and homelessness must think and write about these problems. The whole process of education is an attempt to grow through the vicarious experiences we learn from others. Within Roman Catholicism today it is somewhat ironic that married male lay theologians and philosophers in the United States are among the staunchest defenders of existing hierarchical sexual teaching. However, celibate males need to be very self-critical to avoid their own narrow prejudices and perspectives. Without doubt the exclusive role of male celibates in church power and teaching on all levels has colored Catholic sexual ethics.

One example in the history of Catholic social teaching which shows the male bias was the rationale for the condemnation of masturbation. The general rule was that sexuality existed for the purpose of procreation within a permanent and indissoluble marriage. Masturbation was condemned because such sexual actuation frustrated the procreative purpose of sexuality and the male seed. However, the condemnation of masturbation based on the frustration of the procreative purpose of the semen obviously applied only to males. The whole perspective of the teaching was masculine. Later ethicists had to adduce a different rationale to condemn both male and female masturbation. In fact, a proper understanding of female physiology and the function of the clitoris logically would lead to a questioning of the exclusive emphasis on procreation ([36], p. 17). In many ways a patriarchal approach has had very negative influences on the Catholic tradition in sexuality.

Legal Considerations. The theology, ethics, and understanding of sexuality in the Catholic tradition have been greatly affected by the legal concerns of

Catholicism. Marriage and sexuality have always figured prominently in church law since the church has a significant interest in the ordering of sexuality and marriage. In fact, even the new code of canon law promulgated in 1983 devotes more canons to marriage than to any other subject ([10], p. 740).

The golden era of church or canon law began with the monk Gratian who in 1140 compiled all the laws of the church. The twelfth and thirteenth centuries witnessed a remarkable growth and systematization of church law in all areas but especially with regard to sexuality and marriage. At this time marriage was understood primarily as a contract that came about through the free consent of the spouses but only became fully indissoluble when consummated through sexual intercourse. "By the mid-thirteenth century western churchmen had arrived at a fairly clear consensus about the goals of the church's sexual policy, and canon law had devised workable solutions to many of the commonest difficulties" ([3], p. 485). There was not much innovation and development of the basic understanding of marriage after that time.

As time went on customary, royal, and secular law grew and played an evermore active role in the ordering of sexuality and marriage. However, canon law still governs the understanding and practice of marriage in the Catholic Church. Marriage is entered into in accord with that law and judges decide cases about the validity of marriages on the basis of these canons.

Catholic ethics ever since the fifteenth and sixteenth centuries has been closely associated with canon law. Nowhere has the association been closer than in the area of marriage. In the manuals of Catholic moral theology which came into existence in incipient form in the sixteenth century and served as the textbooks of moral theology down to the Second Vatican Council (1962–1965) the ethics of marriage and sexuality have been discussed within the parameters of the canonical understanding.

The canonical understanding of marriage was enshrined in the 1917 code of canon law which has since been replaced by a new law promulgated in 1983. The law which was in effect through most of the twentieth century basically codified what had been in existence for some time. In this legal understanding marriage is a contract brought about by mutual consent which is an act of the will whereby each party grants and accepts a permanent and exclusive right over the body regarding acts which are of themselves apt for the generation of offspring ([21], p. 12).

From a legal perspective, there are great advantages to such an understanding of marriage. A contract is a very significant and important legal instrumentality. Established legal criteria can be used to determine if the contracting parties truly enter into such an agreement. The object of this particular contract understood as acts which are apt for generation can be verified on the basis of visible and external human acts. As in any legal system, deft distinctions were worked out to deal with particular problems. One such distinction concerned the difference between giving the right to these acts and the exercise of that right. In dealing with marriage, Catholic jurists had to make sure that their theories were in accord with the basic approaches of the Catholic tradition. This distinction both recognized as valid the marriage of Mary and Joseph and acknowledged that procreation is the primary end of marriage ([21], pp. 10–15). There are also other positive aspects about the legal understanding of marriage in Catholic law. For example, the understanding of marriage as a consensual contract in the Middle Ages was a great step forward. The freedom and even in some ways the equality of the marriage partners were strongly affirmed by such an understanding ([21], pp. 146–172).

However, from the viewpoint of theological ethics the canonical view of marriage in the Catholic understanding is inadequate. The sacrament of marriage cannot be reduced to a legal contract. Marriage is a live-giving covenant of love. In its legal understanding Roman Catholicism had a theology of only the first day of marriage! Once the contract is properly entered into there is nothing more to consider or talk about. However, marriage is a dynamic reality that calls for continual growth and development and not just a contract made once and for all time.

The canonical understanding of marriage does not even mention love. Love apparently is not part of the canonical essence of marriage. In this legal understanding marriage is the giving of the right to acts which are apt for generation. The canonical approach gives emphasis to what is external, visible, and legally verifiable, but from a theological and ethical perspective the union of two bodies is the sign, symbol, and incarnation of the union of two hearts and two persons. There can be no doubt that the centrality of legal concerns in the Catholic tradition has distorted the meaning and significance of Christian marriage.

The Pastoral Constitution on the Church in the Modern World of the Second Vatican Council proposed a more adequate understanding of marriage. Marriage is an indissoluble partnership for sharing marital life and love. The document drops the older terminology of procreation as the primary end of marriage and maintains that marriage as an institution and marital love have as a natural characteristic the orientation to procreation and nurture. Some saw the conciliar document as changing the understanding of marriage in Roman Catholicism but others interpreted it as merely a pastoral perspective which does not change the older juridical understanding of marriage ([21], pp. 248–327).

The new code of canon law promulgated in 1983 does not break away entirely from the older juridical and legal approach, but some definite steps are taken. The first canon on marriage (1055) says that the matrimonial covenant by which a man and a woman establish between themselves a partnership for the whole of life is by its nature ordered toward the good of the spouses and the procreation and education of offspring. This covenant between baptized persons has been raised by the Lord to the dignity of a sacrament. But the second part of the canon maintains that a matrimonial contract cannot validly exist between persons unless by that very fact it is also a sacrament ([10], p. 730). Thus, the new canon law still does not get away from the understanding of marriage as a contract. Throughout the Catholic tradition these legal concerns and considerations have predominated and in the process distorted the theological and ethical understanding of marriage.

Authoritarian Interventions. The Roman Catholic Church has insisted that the church is a visible community with authoritative leaders. Catholic faith recognizes a distinct pastoral teaching function for pope and bishops in the church. However, many Catholic theologians today insist on a distinction between authority and the abuse of authoritarianism. The abuse has unfortunately too often been present in Catholic life. The Catholic Church has never been more centralized, more defensive, and more authoritarian than it was in the period immediately preceding the Second Vatican Council. At this time there were interventions of the hierarchical teaching office in all areas of faith and morals but especially in ethical questions. *Humani Generis*, the 1950 encyclical of Pope Pius XII, maintained that whenever the pope goes out of his way to speak on a controverted issue it is no longer a matter for free debate among theologians ([35], p. 209). Vatican II fostered a changed understanding of the church and many in the church recognized the legitimacy at times of public theological dissent from some noninfallible teachings [8], [35]. However, a recent Vatican instruction recognizes as legitimate only the private communication from the theologian to ecclesiastical superiors about the personal difficulties that one might have with a specific teaching of the hierarchical magisterium [6].

The development of the Catholic tradition on sexuality has suffered from an authoritarianism which has tended to preserve the status quo and to prevent any necessary development and change. For over three centuries it was common teaching in Catholicism that all direct sins against sexuality involve grave matter. According to the Latin axiom – *in re sexuali non datur parvitas materiae* – in sexual matters there is no parvity of matter. This means that any direct sexual sin, even imperfect sexual actuation or pleasure, involves grave matter. According to the Catholic understanding, three conditions were necessary for grave or mortal sin – grave matter, full or sufficient deliberation, and full or sufficient consent. In popular understanding the teaching was often thought to say that every sin against sexuality is a mortal sin. Only in sexuality did the axiom hold that there is no parvity of matter [20]. In no other area of human existence was this true so that such a teaching once again highlighted sexuality as more dangerous and more important than any other facet of Christian life. Today theologians rightly no longer even discuss this question. Authoritarian interventions prevented any free discussion of this question until the time of the Second Vatican Council. For example, in 1612, Claudius Acquaviva, the general of the Society of Jesus, forbad all Jesuits to question the existence of parvity of matter in sexuality even with regard to directly willed imperfect sexual actuation or sexual pleasure ([9], pp. 42–47). Theologians could not and did not discuss this point which so distorted the ethical understanding of sexuality.

Pope Pius XII (1939–1958) spoke authoritatively on more subjects than any other pope. Many of his teachings were in the area of sexuality. Catholic theology at the time saw its function as heavily involving the explanation and defense not only of papal statements but also of the decisions of the various congregations of the Roman curia. Decrees from the Roman congregations would put an end to all theological speculation. The 1963 book on sexual ethics by John C. Ford and Gerald Kelly well exemplifies such an approach to theology. The index lists 27 different topics on which Pope Pius XII spoke in an official capacity [12].

The Second Vatican Council opened up new thinking and new approaches within Catholicism. However, the council itself showed the negative effects of an authoritarian intervention in the matter of sexuality. Pope Paul VI took the issue of artificial contraception out of the hands of the council and reserved the decision in this matter to himself ([18], p. 63). Sexual ethics is one area where contemporary Catholic hierarchical teaching regularly refers to pre-Vatican II teaching. As a result of this authoritarian intervention, this important area was left unquestioned and undiscussed in circumstances in which such great change came about in other church understandings and approaches.

In the last few years the hierarchical teaching office in the Catholic Church has often taken action against theologians writing in the area of sexuality. In Europe the interventions have dealt with the work of Ambroggio Valsecchi and Stephan Pfürtner [19]. In the United States church authorities have intervened in a number of different cases. Jesuit John McNeill, after his book on *The Church and the Homosexual*, was silenced by church authorities and later dismissed from the Society of Jesus when he broke his silence to criticize the 1986 document of the Congregation for the Doctrine of the Faith on homosexuality. Rome forced the archbishop of Seattle to take away the imprimatur from Philip S. Keane's *Sexual Morality: A Catholic Perspective* which had been published with his imprimatur in 1977. In 1977 a committee of the Catholic Theological Society of America published *Human Sexuality: New Directions in Catholic Thought* which proposed a newer theory and somewhat different conclusions about sexuality. Church authorities not only published observations expressing their disagreements but also took disciplinary action against Anthony Kosnik, the chair of the committee that produced the volume ([15], pp. 247–250).

In 1986 after a seven year investigation, the Congregation for the Doctrine of the Faith concluded that I was neither suitable nor eligible to teach Catholic theology. The reason for this judgment was my nuanced dissent from hierarchical teachings on a variety of sexual issues such as artificial contraception, sterilization, masturbation, divorce, and homosexuality [16]. In the present climate, Catholic ethicists are often steering clear of the area of sexuality because they fear that action might be taken against them by church authorities if they disagree with hierarchical teaching. However, without sustained and disciplined theological dialogue which at times will take the form of dissent the credibility of the hierarchical teaching itself will suffer and the need to develop the teaching will not be recognized.

Natural Law Justification. Catholic neoscholastic theology and philosophy which reigned supreme before the Second Vatican Council developed a natural law theory to explain and justify the hierarchical sexual teachings [7]. The Catholic tradition has generally emphasized that the moral teachings dealing with life in the world are based not primarily on Scripture but on the natural law which is available to all humankind. In general, natural law insists that human reason reflecting on human nature can arrive at moral wisdom and knowledge for the Christian. However, there were two characteristics of Catholic neo-scholastic natural law theory which strongly supported the hierarchical teaching but which have recently been increasingly questioned by many Catholic theologians – classicism and physicalism.

Classicism refers to a perspective that sees reality in terms of the eternal, the immutable, and the unchanging. The universal essence of human nature is the same in all times, places, and cultures. Such an approach insists on immutable absolute norms that are always and everywhere true. This understanding of reality results in a heavily deductive methodology and tends to absolutize the position of a particular period and culture by making it into an unchanging, universal, metaphysical reality. A more historically conscious methodology gives greater importance to the particular, the contingent, the historical, and the changing without, however, embracing sheer existentialism or relativism. Historical consciousness generally employs a more inductive methodology and does not claim to achieve the absolute certitude of the more deductive approaches. Historical consciousness better corresponds with the reality of how sexual norms did develop and evolve in history.

Physicalism refers to the identification of the human and the moral reality with the physical and the biological aspects of the human act. The anthropological basis of physicalism understands the human being as a rational animal. The noun is animality and the rational modifies or adds to it but does not change the basic given animal or biological structures. Thus, for example, sexuality is understood as that which is common to human beings and to all the animals. The primary end of marriage and sexuality is that which is common to both – the procreation and education of offspring. The human being can never interfere with the physical act of sexual intercourse so that artificial contraception is always wrong. But likewise artificial insemination even with a husband's semen is wrong because the physical act of insemination is normative and must always be present.

Catholic neoscholastic thinkers thus developed a natural law theory to explain coherently and defend hierarchical teaching with a claim that such a theory was based on human reason and human nature and thus open to all. However, such a theory is not as universal as its defenders claim precisely because of the particular way they have developed this theory with its classicism and its physicalism.

In my judgment all of these factors have influenced and supported the development of the Catholic tradition and the present hierarchical teaching. However, this is only one side of the story.

POSITIVE ASPECTS OF THE ROMAN CATHOLIC TRADITION

There are positive aspects in the Roman Catholic tradition that can serve as a basis for developing a more adequate sexual ethic. However, the first part of this discussion should not be seen only in a negative light. To be self-critical is itself a rather significant aspect of the Catholic tradition which follows from many of the perspectives which will be developed shortly. It is important to know the negative aspects so that one can correct them and avoid them in the future. From my perspective the positive aspects of the Catholic tradition which will be developed here are more fundamental so that they can serve to correct and overcome the negative aspects. Five positive aspects of the Catholic tradition will be considered – a positive assessment of the human, the importance of reason, the role of tradition, a communitarian approach, and a non-authoritarian understanding of the teaching role in the church. These five aspects will now be discussed.

A Positive Assessment of the Human. The Catholic tradition has insisted on a positive appreciation of the human. A fundamental question for all religions concerns how the divine and the human relate to one another. Salvation in the Catholic tradition has been understood as bringing the human to its own perfection and not a denial or repression of the human. The treatment of human beings in the *Summa Theologiae* of Thomas Aquinas (Ia IIae, q.1–5) begins with a discussion of the ultimate end of human beings which is happiness or the basic fulfillment of their nature. The two fundamental drives in the human being are the search for truth and the quest for the good. No created or finite reality can ever fully satisfy these two basic drives. God alone is the ultimate end of human beings because in the beatific vision our intellect will know perfect truth and our will will love perfect good. Many Catholic thinkers today would disagree with aspects of the Thomistic understanding, but the Catholic tradition insists that salvation involves the fulfillment and ultimate happiness of the human being.

In the same manner as salvation, grace in the Roman Catholic tradition was never seen as opposed to the human. An old scholastic and spiritual axiom recognized that grace does not destroy nature but grace builds on nature. Too often in the recent past, such an axiom was understood as endorsing a two-story anthropology and understanding of reality – the realm of the natural with the realm of grace or the supernatural on top of it. But for our purposes the axiom indicates the positive relationship between human nature and grace.

The history of theology shows frequent shifts from a more transcendent to a more immanent understanding of God and vice versa. The problem at times seems to be that if one gives a greater role to God, one must give a lesser role to the human or vice versa. At its best the Catholic tradition has avoided this pseudo problem. An ancient patristic adage asserted that the glory of God is the human person come alive. God's glory and human fulfillment are not opposed. The traditional Catholic solution to this question involves the recognition of mediation and participation which are very central in Catholic theology and influence the whole understanding of the divine-human relationship. Mediation avoids the need to choose between either God or the human. The human shares and participates in the goodness and the power of God. By attributing more to the human one does not take away from the divine. God and the human are not in competition. Thus, the Catholic understanding has traditionally seen the human as something good which through grace and salvation is brought to its fulfillment and perfection.

In the case of ethics or moral theology the Catholic tradition has insisted on the goodness of the human and of human reason in arriving at moral wisdom and knowledge. The natural law approach has been a distinctive characteristic of Catholic moral methodology and until the recent past distinguished Catholic ethics from many other forms of Protestant ethics [7]. Natural law is a complex theory, but in the first place it was a response to the question: where does the Christian and systematic Christian ethical thinking find moral wisdom and knowledge? Some maintain that only in revelation can the Christian find moral wisdom and knowledge. Natural law recognizes that human reason on the basis of human nature can arrive at true moral wisdom and knowledge. A glance at any textbook of Catholic moral theology in the period before the Second Vatican Council indicates an almost exclusive emphasis on natural law. Appeals to scripture were usually only in the form of prooftexts to corroborate what had already been determined and decided on the basis of natural law. Hierarchical Catholic teaching in all areas including sexual and social ethics emphasized a natural law methodology. As noted above, I have problems with the way human reason and human nature were understood, but this method gave great importance to the role of the human.

The theological basis for the natural law rested on the doctrine of creation. God created the world and saw that it was good. Human beings can discern through their reason the order that God puts into the world and how they as human beings should thus act in accord with God's plan. Unlike some other Christian approaches, the Catholic understanding does not see sin as totally destroying or doing away with the reality of creation.

Catholic moral theology can be criticized for not giving enough importance to the reality of sin and its effects on the human. A natural law optimism has at times affected the Catholic approach precisely because this theory did not give enough importance to the role of sin. I wholeheartedly agree that sin does not destroy the order and reality of creation, but sin has affected it. To overcome the dangers of a one-sided emphasis on the goodness of creation and the human in the Catholic tradition as well as the danger of separating out the order of nature from the order of the supernatural, I have proposed what I call the stance or basic perspective of moral theology. The Christian and Christian ethics understand the world and all in it in light of the fivefold Christian mysteries of creation, sin, incarnation, redemption, and resurrection destiny. All human reality is touched by these mysteries. Whatever exists shows the goodness of creation, has been somewhat affected by sin, but is touched by the incarnation and redemption while always falling short of the fullness of resurrection destiny. I have criticized recent Catholic moral theology and hierarchical church teaching for at times overemphasizing the basic goodness of the human.

The Catholic acceptance of the goodness of the human determined not only the methodology of moral theology but also the positions taken on many important content questions. A very significant question in social ethics concerns the understanding of political authority and political structure. Some Christians have seen political structure and the state as an order of preservation brought about by God to keep sinful human beings from destroying one another. In such a view, the state exists primarily to keep order and to prevent evil from getting out of hand. In the Catholic tradition political society or the state is seen as natural and based on human nature. Human beings are by nature social and called to live in political society, for only in such society can human beings begin to achieve their fulfillment and perfection. The end of political society in such an approach is the positive one of working for justice and the common good [32].

The Catholic tradition in all aspects and especially in ethics has upheld the basic goodness of the human. Such an approach should logically also color the whole attitude and approach to human sexuality. Here, too, I would employ the stance mentioned above. Sexuality as created by God is good; sin can and does affect human sexuality; through the incarnation and redemption sexuality and the sexual relationship are touched by the saving love of God, but sexuality like all human reality will be transformed in the new heaven and the new earth.

The Catholic emphasis on mediation forms the basis for Catholic insistence on sacramentality [17]. Sacramentality can be understood in a very broad and general sense. Here everything that has been created comes forth from the hand of God and shows forth and reveals the creator. Likewise, redemption is mediated and revealed in and through the human and human history. Thus all that exists in the world is a sacrament or sign that reveals the presence and work of God. All creation shows forth the glory of God. The Catholic tradition has gone so far as to claim that reason can prove the existence of God by going from creation to the creator. One can still uphold this basic Catholic sacramental vision based on human mediation of the divine without necessarily claiming that one can conclusively prove the existence of God on the basis of reason alone.

The sacramental understanding of all reality serves as the foundation for the role of the sacraments in Catholic life and worship. The fundamental reality is that God is mediated in and through the human. Jesus is the primary sacrament of God, for the humanity of Jesus makes visible the saving power and presence of God. The church or the community of disciples of Jesus is the sign that makes present the reality of the risen Jesus at work in the world. The seven sacraments are the privileged signs by which this community celebrates the presence of the risen Jesus in our midst. Basic human realities such as a celebratory meal become the way in which the church celebrates the saving presence of God in our midst. The human reality of marriage has been recognized as one of the seven specific sacraments. Thus, the marital and sexual union of husband and wife become the sign and symbol of the union and fruitful love of God for God's people. Such a sacramental understanding can never disparage the human in general and in this case, human sexuality. Joan Timmerman has made sacramentality the primary basis for her theological ethic of sexuality ([36], pp. 1–9). Thus, the basic acceptance of the human is a very fundamental tenet of Catholicism and must influence the way sexuality is understood. The negative dualisms of the tradition must be corrected in the light of this more basic and fundamental vision.

Faith and Reason. The Catholic emphasis on the goodness of the human is expressed in the epistemological order by the insistence that faith and reason cannot contradict one another. The Catholic tradition has warmly embraced this recognition of the role of reason and its fundamental compatibility with faith. Such an understanding in its own way is a great act of faith in the power or reason. History shows that Catholicism has not always lived up to this theoretical understanding, but the basic principle has been a cornerstone of Catholic theology.

Theology has played a significant and important role in Catholic life, for theology is necessary for the continued existence of the community of faith which must always try to understand, live, and appropriate in a self-critical way the word and work of Jesus. Theology has been understood as faith seeking understanding and understanding seeking faith – a strong assertion that faith and reason are not at enmity with one another.

The compatibility between faith and reason can be seen in many different ways. The first universities came into existence under the auspices of the church. The church fostered the search for truth and was not afraid of reason. The earliest empirical scientists were themselves theologians and philosophers who were committed to finding out the truth.

The history of Catholic theory and practice shows the fundamental compatibility between faith and reason in this tradition. The early church borrowed heavily from Greek philosophy in its attempt to understand better the Christian faith and message. Thomas Aquinas (d. 1274) ranks as the most outstanding theologian in the Catholic tradition. The genius of Aquinas was to use the Aristotelian thought which was just then coming back into the European university world to interpret and explain better the Catholic faith. From a more practical perspective, Catholicism took over the reality of Roman law and employed it for the governance of the church.

History shows there have been many tensions between faith and reason. Often the Catholic Church has lost the nerve of its medieval theologians who boldly asserted that faith and reason cannot contradict one another. Aberrations have led the popular imagination to see an insuperable gulf between faith and empirical sciences. However, in theory Catholicism has always been open to the insights of human reason and has found truth in and through the use of human reason.

Two points are most relevant for the concerns of this paper. Catholic approaches must learn from whatever the human sciences can teach us about human sexuality. Faith does not supply for the defects of reason. Yes, every science is limited and no one empirical science can ever be totally identified with the human, but the sciences can furnish important knowledge about human sexuality.

Second, Catholicism at times has been so supportive of human reason that it has tended to become too overly rational and not give enough importance to nonconceptual knowledge and to human inclinations and feelings. However, the same basic theological understanding that supports reason also supports a positive role for nonconceptual knowledge as well as for human inclinations and feelings. These are all important sources for a proper understanding of human sexuality. In theory Roman Catholicism is thus open to search for and find the meaning of sexuality in and through all aspects of human experience. In theory Catholic theological ethics has at its disposal everything that can contribute to a better understanding and appreciation of the meaning of human sexuality and the values and principles that should guide its use.

Scripture and Tradition. The Catholic emphasis on scripture and tradition also equips Catholic thought at the present time with perspectives that are conducive to a more adequate understanding of human sexuality. I do not agree with all the ways in which scripture and tradition have been understood in the Catholic approach, but the basic thrust of this understanding provides helpful perspectives for one trying to develop an adequate theology and ethics of sexuality.

The recognition of the need for scripture and tradition has often been contrasted with an approach emphasizing the scripture alone. With this perspective the Catholic approach has always been able to avoid a biblical fundamentalism. The word and work of Jesus must always be understood and appropriated in the light of the ongoing historical and cultural realities in which we live. The Catholic approach has never been content merely to repeat what has been found in the scripture. Revelation has not been identified only with the words of scripture. Catholicism at times has not given enough importance to scripture, but biblical fundamentalism has never been a Catholic problem.

It is quite ironic that the method of feminist biblical criticism has a basic compatibility with the traditional Catholic approach. Catholicism has always been willing to bring other presuppositions to bear on how one interprets the Bible.

A proper understanding of tradition does not see it as something that has ceased but as an ongoing attempt to understand the word and work of God in changing circumstances [26]. The church must constantly be striving to know and live better the meaning of human sexuality in light of the changing conditions of time and place.

One negative possibility with the Catholic understanding of tradition is to identify what has always been as a true divine tradition revealing the will of God. Thus, for example, official Catholic hierarchical teaching claims that the historical reality of the church's never having ordained women is proof of the divine will [5]. However, the Catholic recognition that God works in and through the human can serve as the basis for saying that the tradition of not ordaining women has depended upon the sociological circumstances of time and place and does not represent the eternal and immutable will of God.

By taking tradition seriously the Catholic church has also paid great attention to its own historical development. Historical studies have always played an important role in the Catholic tradition. The renewal of the Second Vatican Council was heavily influenced by a return to the sources. The existing realities of the twentieth century were criticized in the light of what had existed in earlier periods in the life of the church. Here too, there can be a danger of merely absolutizing what is past or of not giving enough importance to the signs of the times, but traditions and history have always been studied and appreciated in Roman Catholicism.

Such an interest in historical development should be helpful in trying to determine how certain understandings were arrived at and how particular values and norms came to the fore. The critical study of history can also unmask some of the shortcomings and problems that have existed in the past and even continue today.

Recently many significant monographs have been published dealing with historical developments with regard to matters of sexuality and gender. Some of these studies have raised up dangerous memories that have been forgotten or were purposely erased from the living memory of the church. But an interest in tradition and historical development is of great importance to our learning about these developing realities. Perhaps most significant here are studies about the role of women in the church. For example, the role of women in the medieval church has been the subject of much recent scholarship [28], [34].

In trying to discern the present it has helped to know how the past came about. John T. Noonan, for example, has done an exhaustive history of the Catholic teaching on contraception. On the basis of such a history, Noonan concluded that Roman Catholicism could and should change its teaching condemning artificial contraception for spouses [25]. Noonan had earlier studied a change of the teaching on usury in Catholicism and saw parallels for a similar change in the teaching on contraception [24]. Important recent historical studies include Peter Brown's work on sexuality and the body in the early church [2] and James Brundage's *Law, Sex, and Christian Society in Medieval Europe* [3]. John Boswell has studied gay people in western Europe from the beginning of the Christian era to the fourteenth century [1]. Many historical studies have been done on Christian marriage which point out how various developments occurred [21]. Such historical studies help to form a more critical appreciation of the tradition and how it has developed. This critical understanding and appreciation of the past can assist in developing a more adequate sexual ethic for today.

A Communitarian Approach. On the American scene the danger of individualism continues to be present. This affects our approach to many questions including sexuality. In my judgment the primary source of the criticisms made by the United States Catholic bishops about the United States economy in their recent pastoral letter stems from the individualism so prevalent in American life [23]. The danger is that sexuality will also be seen in too narrow a perspective concerned only with the individual. The Catholic tradition in general and in its understanding of sexuality has always avoided a narrowly individualistic perspective ([4], pp. 139–143). The communitarian approach in Catholicism emphasized that sexual relations involve a committed relationship of two people which also has a concern for the broader society. The Catholic tradition has insisted that sexuality must always be related not only to the individuals involved but also to the good of the species. Ironically, one can and at times should use the relationship to the species to argue for limits on the number of children to be had in certain situations. Historical studies such as Brown's and contemporary feminist concerns also insist on the social aspect of sexual ethics ([36], pp. 50–68). Yes, there are important questions about how the communitarian and social aspects of sexuality influence our judgments and our governing values and norms, but especially on the contemporary American scene one must be aware of the danger of a too individualistic approach.

A Renewed Understanding of the Hierarchical Teaching Role. At the present time in Roman Catholicism the issue of sexuality has become predominantly a question of ecclesiology. The primary concern is how the hierarchical teaching office carries out its function. This teaching office has been unwilling to change any of its positions and also has strongly denied the legitimacy of public theological dissent as well as dissent in practice from existing norms. Thus, the ecclesiological aspect has become central.

This is not the place to develop an ecclesiology which calls for a less authoritarian understanding of the function of the hierarchical teaching office and recognizes the legitimacy of some dissent on these issues within the church. Many of the existing controversies in Catholicism center on questions of authority and frequently include sexual issues. It is sufficient to point out that less authoritarian approaches to the hierarchical teaching office have been proposed, but in practice the pope and bishops have rejected such understandings [8], [22], [27]. It is not going out very far on a limb to assert that contemporary Catholicism will continue to experience tensions on these questions in the immediate future.

The focus of this paper has been on Catholic theology and ethics in dealing with sexuality. In the long term, I think the Catholic theological and ethical traditions have the tools to develop a more adequate sexual ethic. But that takes time. Meanwhile, individual Catholics must make their own decisions and many people are still suffering ill effects because of their struggle with the existing hierarchical teaching. What can be done now?

The positive aspects of the Catholic tradition such as the basic goodness of the human and of human reason point toward another distinctive characteristic of Catholic morality which is very significant – its intrinsic nature. The Thomistic tradition which has been so important in Roman Catholicism insists that morality is intrinsic – something is commanded because it is good and not the other way around. Morality involves the true fulfillment and happiness of the person. Catholic moral theology has even been criticized for giving too much attention to self-fulfillment and a eudaimonistic approach.

Such an understanding gives a very positive role to the human person in moral decision making. However, this does not mean that anything goes. The dilemma of the moral life can easily be summarized – one must follow one's conscience, but conscience might be wrong. From a theological perspective human finitude and sinfulness are the sources of possible errors in conscientious decision-making. All of us are limited and not able to see all that is to be seen. All are sinful and at times capable of self-deception. The prudent decision maker must recognize the inherent limitations of conscience and try to overcome them. There are no short cuts or easy answers, but in the end one must be true to oneself and one's conscience. For persons who have been warped and negatively affected by aspects of the Catholic tradition, clinicians can help them to come to this more mature understanding of conscience which is in accord with the best of the Catholic tradition.

Roman Catholicism like many individuals and communities today is struggling to come up with a better understanding of human sexuality and the morality that should govern it.

The struggle will not be easy and will never be fully achieved. A reflective and self-critical analysis shows many of the negative aspects in the Catholic tradition's attempt to deal with sexuality in the past. However, this paper also points out some fundamental perspectives in the Catholic tradition which should help it to correct the problems and develop a more adequate sexual ethic.

Southern Methodist University
Dallas, Texas, U.S.A.

BIBLIOGRAPHY

1. Boswell, J.: 1980, *Christianity, Social Tolerance, and Homosexuality,* University of Chicago Press, Chicago.
2. Brown, P.: 1988, *The Body and Society: Men, Women, and Sexual Renunciation in Early Christianity,* Columbia University Press, New York.
3. Brundage, J.A.: 1987, *Law, Sex, and Christian Society in Medieval Europe,* University of Chicago Press, Chicago.
4. Cahill, L.S.: 1985, *Between the Sexes: Foundations for a Christian Ethics of Sexuality,* Paulist, New York.
5. Congregation for the Doctrine of the Faith: 1977, *Declaration on the Question of the Admission of Women to the Ministerial Priesthood,* United States Catholic Conference, Washington.
6. Congregation for the Doctrine of the Faith: 1990, 'Instruction on the Ecclesial Vocation of the Theologian', *Origins* 20, 117–126.

7. Curran, C.E. and McCormick, R.A. (eds.): 1991, *Readings in Moral Theology No. 7: Natural Law and Theology,* Paulist, New York.
8. Curran, C.E. and McCormick, R.A. (eds.): 1988, *Readings in Moral Theology No. 6: Dissent in the Church,* Paulist, New York.
9. Diaz Moreno, J.M.: 1960, 'La Doctrina Moral sobre la Parvedad de Materia in Re Venerea desde Cayetano hasta S. Alfonso', *Archivio Teológico Granadino* 23, 5–138.
10. Doyle, T.P.: 1985, 'Marriage', in J. Coriden et al. (eds.), *The Code of Canon Law: A Text and Commentary,* Paulist, New York, pp. 737–833.
11. Erler, M. and Kowaleski, M. (eds.): 1988, *Women and Power in the Middle Ages,* University of Georgia Press, Athens, GA.
12. Ford, J.C. and Kelly, G.: 1963, *Contemporary Moral Theology II: Marriage Questions,* Newman Press, Westminster, MD.
13. Greeley, A.M.: 1976, 'The Lay Reaction', in H. Küng and L. Swidler (eds.), *The Church in Anguish: Has the Vatican Betrayed Vatican II?,* Harper and Row, San Francisco.
14. Greeley, A.M. et al.: 1976, *Catholic Schools in a Declining Church,* Sheed and Ward, Kansas City, MO.
15. Griffin, L.: 'American Catholic Sexual Ethics', in S. Vicchio and V. Geiger (eds.), *Perspectives in the American Catholic Church 1789–1989,* Christian Classics, Westminster, MD, pp. 231–252.
16. Häring, B.: 1988, 'The Curran Case: Conflict between Rome and the Moral Theologian', in H. Küng and L. Swidler (eds.), *The Church in Anguish: Has the Vatican Betrayed Vatican II?,* Harper and Row, San Francisco, pp. 235–250.
17. Irwin, K.W.: 1987, 'Sacrament', in J. Komonchak et al. (eds.), *The New Dictionary of Theology,* Glazier, Wilmington, DE, pp. 910–922.
18. Kaiser, R.B.: 1985, *The Politics of Sex and Religion: A Case History in the Development of Doctrine, 1963–1984,* Leaven, Kansas City, MO.
19. Kaufmann, L.: 1987, *Ein Ungelöster Kirchenkonflikt: Der Fall Pfürtner,* Edition Exodus, Freiburg, Switzerland.
20. Kleber, H.: 1971, *De Parvitate Materiae in Sexto,* Pustet, Regensburg.
21. Mackin, T.: 1982, *What is Marriage?,* Paulist Press, Mahwah, NJ.
22. May, W.W. (ed.): 1987, *Vatican Authority and American Catholic Dissent,* Crossroad, New York.
23. National Conference of Catholic Bishops: 1986, *Economic Justice For All: Pastoral Letter on Catholic Social Teaching and the U.S. Economy,* United States Catholic Conference, Washington.
24. Noonan, J.T.: 1957, *The Scholastic Analysis of Usury,* Harvard University Press, Cambridge, MA.
25. Noonan, J.T.: 1986, *Contraception: A History of its Treatment by the Catholic Theologians and Canonists,* enl. ed., Harvard University Press, Cambridge, MA.
26. O'Malley, J.W.: 1989, *Tradition and Transition: Historical Perspectives on Vatican II,* Glazier, Wilmington, DE.
27. Penaskovic, R. (ed.): 1987, *Theology and Authority,* Hendrickson, Peabody, MA.
28. Petroff, E.A. (ed.): 1986, *Medieval Women's Visionary Literature,* Oxford, New York.
29. Pope John Paul II: 1981, *Familiaris Consortio: The Role of the Christian Family in the Modern World,* St. Paul Editions, Boston.
30. Pope Pius XI: 1957, 'Casti Connubii', in T.P. McLaughlin (ed.), *The Church and the Reconstruction of the Modern World: The Social Encyclicals of Pius XI,* Doubleday Image, Garden City, NY, pp. 118–170.
31. Quinn, J.R.: 1980, 'New Context for Contraception Teaching', *Origins* 10, 263–267.
32. Rommen, H.: 1945, *The State in Catholic Thought,* B. Herder, St. Louis.
33. Ruether, R.R.: 1987, *Contemporary Roman Catholicism: Crises and Challenges,* Sheed and Ward, Kansas City, MO.
34. Ruether, R.R. (ed.): 1974, *Religion and Sexism: Images of Women in the Jewish and*

Christian Traditions, Simon and Schuster, New York.
35. Sullivan, F.A.: 1983, *Magisterium: Teaching Authority in the Catholic Church,* Paulist Press, New York.
36. Timmerman, J.: 1986, *The Mardi Gras Syndrome: Rethinking Christian Sexuality,* Crossroad, New York.

JAMES B. NELSON

BODY THEOLOGY AND HUMAN SEXUALITY

It is commonly observed that religion is a very ambiguous human enterprise. The creative power of religion is great, for the divine presence is, indeed, often mediated with life-giving power through religious patterns of doctrine, morals, worship, and spirituality. The religious enterprise is also one of the most dangerous of all human enterprises, since it is always tempted to claim ultimate sanction for its humanly-constructed beliefs and practices. Nowhere is this mix of the creative and the destructive, this painful ambiguity, more apparent than in the arena of human sexuality. Nowhere is it more apparent, because sexuality's own creative and destructive capacities are such that religion has given an unusual amount of attention to this dimension of human life, has attempted to control it, and has shown considerable fear of it.

SEVEN SINS AND SEVEN VIRTUES

Early in Christian history two lists arose: the seven deadly sins and the seven virtues [18].[1] Though I here make no attempt to recapitulate those early lists, I want to name seven deadly sins through which the Western religious traditions have contributed to sexual "dis-ease," countered by seven virtues or positive resources which the traditions offer to nurture sexual health. My assumptions are twofold. First, the sexual distortions in these religious traditions have largely resulted from perversions of their own central teachings; hence, through reclaiming that which is more authentic to the core of these faiths there may be sexual healing. Second, each of these seven sins betrays profound suspicions of the human body; thus the need for doing body theology is immense (a subject which the later sections of this chapter will explore).

Spiritualistic Dualism Is the First Deadly Sin. With its counterpart, sexist or patriarchal dualism, spiritualistic dualism underlies and gives shaping power to all the other sins of the list. Any dualism is the radical breaking apart of two elements which essentially belong together, a rupture which sees the two coexisting in uneasy truce or in open warfare.

Though quite foreign to Jewish scriptures and practice, spiritualistic dualism was grounded in Hellenistic Greco-Roman culture and had a profound impact on the early Christian church. Continuing with power to the present, it sees life composed of two antagonistic elements: spirit, which is good and eternal, and flesh or matter, which is temporal, corruptible and corrupting. The religious enterprise, then, is essentially the escape from the snares of bodily life through

Ronald M. Green (ed.), Religion and Sexual Health, 37–54.

the spirit's control. The sexual body is believed to be the particular locus of human sin.

However, there is a countervailing virtue in both Judaism and Christianity, much more authentic to the roots of each faith. In Judaism it is a strong doctrine of the goodness of creation and with it an anthropology that proclaims the unity and goodness of the human body-self. The Song of Solomon is a lyrical celebration of the delights of unashamed fleshly love.

Christianity also expresses this anti-dualistic virtue in affirming creation, and couples this with its particular emphasis on divine incarnation. Incarnation: the most basic and decisive experience of God comes not in abstract doctrine or mystical otherworldly experiences but *in flesh.*True, the faith's ongoing struggle with dualism has been evident in its marked discomfort over taking incarnation radically. Christian doctrine and piety have been strongly inclined to confine the incarnation of God to one and only one human figure – Jesus of Nazareth. And persisting body denial has made most Christians qualify that humanity through their silence about or actual denial of Jesus's sexuality.

There is another possibility however implausible it may seem to some: without denigrating the significance of God's revelation in Jesus, incarnation might yet be understood more inclusively. Then the fleshly experience of each of us becomes vitally important to our experience of God. Then the fully physical, sweating, lubricating, menstruating, ejaculating, urinating, defecating bodies that we are – in sickness and in health – are the central vehicles of God's embodiment in the world.

Yet, the Christian suspicion of the body and its pleasures continues, and the sexual purity campaigns did not end with the Victorian era. But at the authentic core of both religious traditions is the conviction that sexuality and spirituality are not antithetical. The creation-affirming Jewish faith and the incarnationalist Christian faith attest to the goodness of the body-self with all of its rich sexuality as intrinsic to God's invitation into our full humanness and loving communion.

The Second Deadly Sin Is Sexist or Patriarchal Dualism. This systematic and systemic subordination of women is the counterpart of spiritualistic dualism, for men typically have defined themselves as essentially spirit or mind and have defined women as essentially body and emotion. The logic, of course, is that the higher reality must dominate and control the lower.

Patriarchal dualism pervades both Jewish and Christian scriptures and cultures. In Christianity, however, it has taken particular twists that have powerfully joined the male control of women to body-denial. Major and classic understandings of the atonement, Christ's suffering on the cross, have undergirded attitudes that accept and encourage suffering. Concomitantly, Christian theology has often justified both the deprivation of sensual pleasure and the acceptance (even, at times, the seeking) of pain as intrinsic to the attainment of salvation's joys. But *women's* suffering has particularly been undergirded, for in patriarchy it is they and not males who essentially represent the evil that needs redemption – the fleshly body [3]. That sexist dualism is a deadly sin to the health and well-being of women needs no elaboration. That it is also enormously destructive for males, even while men continue to exercise dominant social power and privilege, needs to be recognized as well.

The good news, the countervailing virtue in these religious traditions, is the affirmation of human equality. In his better moments the Apostle Paul wrote, "There is neither Jew nor Greek, there is neither slave nor free, there is neither male nor female, for you are all one in Christ Jesus" (Galatians 3:28). The second great wave of feminism in our society, occurring in the latter third of the current century, has produced real gains in gender justice and inclusiveness – few would doubt this. That Jewish and Christian communities have far to go is also beyond question. Continuing resistance to women's religious leadership and ongoing religious justifications for male control of women's bodies are but two of many sad illustrations possible.

Nevertheless, gender equality is a truer expression of the heart of our common religious heritage. It is sexism that is the religious perversion. At the same time, it is apparent that the continuing power of sexist dualism manifests the fear of the body – the body that appears essential to woman's being in a way that it is not for the man. All of our body issues are enormously complicated by the interplay of these two faces of dualism, as are all of the major moral issues of our day. Not only are the more obvious issues of body rejection, sexism, homophobia and heterosexism rooted in dualistic dynamics, but so also are many of the dynamics of social violence, militarism, racism, economic oppression, and ecological abuse.

The Third Deadly Sin Is Heterosexism (Socially-Enforced Compulsory Heterosexuality) and its Companion Phenomenon Homophobia (the Irrational Fear of Same-Sex Feelings and Expression). Tragically, this sin has pervaded both Jewish and Christian histories. Yet, it cannot be justified by careful biblical interpretation. The bible does not even deal with homosexuality as a psychosexual orientation, an understanding that did not arise until the 19th century. While scriptures do condemn certain homosexual expressions, they appear largely to disapprove of these acts because of the lust, rape, idolatry, violation of religious purity obligations, or the pederasty expressed in these specific contexts. There is no explicit biblical guidance on same-sex genital expression within a context of mutual respect and love – though there are biblical celebrations of same-sex emotional intimacy, a fact often overlooked by the fearful proof-texters.

The dynamics of homophobia are numerous and complex. Frequently they are deeply rooted in misogyny, in the fear of and contempt for the "failed male," in the fear of one's own bisexual capacities, in general erotophobia (the fear of sexuality itself), and in the alienation from one's own body and hence the desperate envy of anyone who appears to be more sexual than oneself.

The good news – the virtue – is that Jews and Christians have significant resources for dealing with these things. Those religious convictions that resist dualisms also undercut heterosexism and homophobia. Further, central to each faith is God's radical affirmation of every person, each unique body-self. And experience of grace is the foundation of a deep sense of personal security that releases us from the anxious need to punish those who seem sexually different from ourselves. Then the issue becomes not that of sexual orientation as such, but is rather whether, whatever our orientation, we can express our sexuality in life-giving ways.

That both faith communions are making some progress on issues of sexual orientation seems evident from a number of indices. That the subject is currently the most divisive one for many congregations and judicatories is also evident, as witness the intensity of the debate over lesbian and gay ordination. In all of this one fact seems clear: fear of the body is a central dynamic of resistance to equality in sexual orientation.

The Fourth Deadly Sin Contributing to Sexual Disease Is Guilt over Self-Love. Christian theologies and pieties have had a more difficult time with this than have Jewish. Dominant Christian interpretations all too frequently have understood self-love as equivalent to egocentrism, selfishness, and narcissism, hence incompatible with the religious life. A sharp disjunction has been drawn between agape (selfless, self-giving love believed normative for the faithful) and eros (the desire for fulfillment).

When suspicion about self-love is combined with a suspicion of the body and of sexual feelings, the stage is set for sexual dis-ease. Masturbation is a case in point. To be sure, this subject is no longer inflamed by passions akin to those of the 19th century sexual purity reformers; Sylvester Graham's "graham crackers" and John Kellogg's corn flakes are no longer persuasive as bland diets to prevent the solitary vice. Yet, masturbation is still an obvious arena of guilt, simply because giving oneself sexual pleasure seems to be sheer self-centeredness. But, in a larger sense, self-love has had a bad press, particularly in Christianity. And when self-love is denigrated, authentic intimacy with a sexual partner is made more problematic, for true intimacy always is rooted in the solid sense of identity and self-worth of each of the partners.

The good news is that self-love is *not* a deadly sin. Both Hebrew and Christian scriptures bid us to love our neighbors *as* ourselves, not *instead of* ourselves. Both religious traditions at their best know that love is invisible and nonquantifiable. It is not true that the more love we save for ourselves the less we have for others. Authentic self-love is not a grasping selfishness – which actually betrays the lack of self-love. Rather, it is a deep self-acceptance which comes through the affirmation of one's own graciously given worth and creaturely fineness, our warts and all.

Better theological work in recent decades has brought corrections in earlier simplistic condemnations of self-love. Such theological shifts undoubtedly have been undergirded by a growing psychological sophistication within religious communities. Even more important, Christian and Jewish feminists, gay men, and lesbians have shown how dominant males have made the virtue of self-denial a means of controlling those whose sexuality was different from theirs.

But the battle about self-love is far from over, particularly in its sexual expressions. While theological treatises are beginning to give sexual pleasure some justification, most congregations would still be embarrassed by its open endorsement except, perhaps, a discrete hint during a wedding service. The affirmation of masturbation as a positive good for persons of all ages, partnered or unpartnered, is rarely found in religious writings and even more rarely mentioned aloud in synagogue or church. The sexual and body aspects of self-love surely are not the only dimensions, but they are barometers that remind us how our problems with genuine self-love appear intricately intertwined with our continuing bodily fear.

The Fifth Deadly Sin Is a Legalistic Sexual Ethics. Many adherents of both Christian and Jewish faiths have fallen into more legalism about sexual morality than about any other arena of human behavior. Legalism is the attempt to apply precise rules and objective standards to whole classes of actions without regard to their unique contexts or the meanings those acts have to particular persons. Masturbation, homosexual expression, and non-marital heterosexual intercourse are frequent targets for religio-moral absolutes. So also, however, are numerous issues related to reproduction – contraception, abortion, and various reproductive technologies.

The virtue that speaks to the deadly sin of legalism is *love*. Our body-selves are intended to express the language of love. Our sexuality is God's way of calling us into communion with others through our need to reach out, to touch, to embrace – emotionally, intellectually, and physically. Since we have been created with the desire for communion, the positive moral claim upon us is that we become in fact what essentially we are: lovers, in the richest and deepest sense of that good word.

A sexual ethics grounded in love need not be devoid of clearly-articulated values and sturdy guiding norms. Indeed, such are vitally important. But the morality of sexual expression cannot adequately be measured by the physiological contours of certain types of acts. For example, religious legalism typically has condemned genital sex outside of heterosexual marriage and has blessed sex *within* marriage. Such a morality consequently has prevented us from blessing the loving unions of same-sex couples or finding ways to affirm committed heterosexual relationships short of legal marriage. At the same time, that morality (even if perhaps unwittingly) has given moral justification for unloving and exploitive sex within marriages by insisting that the rightness of sex is measured not fundamentally by the quality of the relationship but by its form.

The alternative to sexual legalism is not laxity and license, but ethics grounded in the centrality of love. Such an ethics is based on the conviction that human sexuality finds its intended and richest expression in the kind of love that enriches the humanity of persons and expresses faithfulness to God. Such an ethics cannot guarantee freedom from mistakes in the sexual life, but it aims to serve and not to inhibit the maturation and human "becoming" of sexual persons.

Many in both Jewish and Christian communions are more open to a non-legalistic approach to sexual ethics than perhaps ever before. But sexual legalism is not a thing of the past. The unbending stringency of Orthodox Judaism, the official Roman Catholic retreat from Vatican II sexuality positions, and the strident moralisms of fundamentalist Protestants are also evident. What is seldom acknowledged and needs to be noted, however, is that religious legalism is much more commonly applied to sexuality and body issues than to any other area of human morality. That should not surprise us. The body is still a great source of anxiety, and we typically want desperately to control that which we fear.

The Sixth Deadly Sin of Which Our Religious Traditions Are Often Guilty Is a Sexless Image of Spirituality. This has been a bane of Christianity more than of Judaism, for the church more than the synagogue has been influenced by the Neoplatonic split between spirit and body. In its more extreme forms, true spirituality was perceived as sexless, celibacy was meritorious, and bodily mortification and pain were conducive to spiritual purification. H.L. Mencken was wrong in his quip about the Puritans. They were not really the ones tortured by the haunting fear that others somewhere else might be enjoying themselves. The Puritans were not that antisexual, but many other Christians were, and some still are.

Good news comes in the recognition that a sensuous, body-embracing, sexual spirituality is more authentic to both Jewish and Christian heritages. We are beginning to see that repressed sexuality "keeps the gods at bay" and does not bode well for the fullest, healthiest spirituality. We are beginning to recognize that the erotic and bodily hungers celebrated in the Song of Solomon are human sharings in the passionate longings of God, the divine One who is shamelessly the earth's Lover. If this conviction is resisted by some religious people, as shown in recent years by their distress over the film *The Last Temptation of Christ,* many others would endorse Nikos Kazantzakis's words from another of his books: "Within me even the most metaphysical problem takes on a warm physical body which smells of sea, soil, and human sweat. The Word, in order to touch me, must become warm flesh. Only then do I understand – when I can smell, see, and touch ([15], p. 43)." Truly, the body remains a central spirituality issue – and a problematic one.

The Seventh Deadly Sin of Our Religions Has Been the Privatization of Sexuality. The word play is intentional. Sexuality has been religiously consigned to the non-public world and narrowed to a genital matter – "the privates." To that extent, the public, institutional, and justice dimensions of human sexuality have often been neglected.

Yet, one of the ironies of American history is that the 19th century "sexual purity movements" most determined to push sex back into the confines and privacy of the marital bed frequently heightened its visibility and made sex a matter of more public discussion. Thus, even Anthony Comstock's war on obscenity unwittingly served Margaret Sanger's movement for birth control.

However, "the personal *is* public." This has become a rallying cry of current feminism, and rightly so. It is also a conviction of the Jewish and Christian religious traditions at their best. Sexuality issues are inevitably political, and the most deeply personal is at the same time connected with the social.

Yet, there are different ways of understanding this. The radical religious right wing of Christianity exemplifies one. Clearly, sexuality issues are at the core of its public agenda: opposition to gender equality, sex education, abortion, homosexuality, pornography, the Equal Rights Amendment, and family planning, on the one hand, and support of those programs that would strengthen "the traditional family," on the other. Yet, for all of its public emphasis on sexuality, the radical right exhibits a thinly-veiled fear of it. The two familiar dualisms shape its agenda: patriarchy's hierarchical ordering of the sexes and spiritualism's denigration of the body. The message becomes clear. The right wing's current public sexual agenda is to get sexuality *out* of the public and back into the private sphere once again. And the private sphere is that of the male-controlled "traditional family."

There is a different way of seeing sexuality as a public issue. It is to recognize the basic falsity of the sharp disjunction between private and public, for it is a dualism deeply related to the sexual dualisms. It is to see that sexual politics is inevitable, for politics (as Aristotle taught us) is the art of creating community, and human sexuality at its core deals with those intimate relationships that shape the larger communities of life. Thus, the bedroom cannot be confined to the bedroom. Justice issues for the sexually oppressed, sexual abuse, reproductive choice, population control, exploitation in commercialized sex, adequate sexuality education – these among others are now obviously public issues. Yet, we are only beginning to understand that there are important sexual dimensions to other vast social issues: racism, social violence, militarism, economic exploitation, environmental abuse. Perhaps we are late in recognizing the sexuality embedded in these latter issues because of our continuing penchant for dualisms. Body anxiety still bids to keep sex private – or to try to return it to the realm of the private – but it will not be so contained.

So, the seven (or more) deadly sexual sins are still very much with us. They are neither the last nor the truest word about our religious traditions. But their persistence signals the need for more adequate body theology.

ON DOING BODY THEOLOGY

Before the last two decades, by far the bulk of Christian and Jewish writings about body and sexuality had assumed that the question was essentially a one-way matter: "what does our *religion* (or the scripture or the tradition or the religious authorities) say about the body and its sexuality?" The assumption was that religion had its truth, received or arrived at quite independently of our bodily-sexual experience, which then needed only to be applied. Faith provided the instruction book which came with the body appliance, an instruction that often seemed to say "CAUTION, READ CAREFULLY BEFORE OPERATING!"

Unfortunately, more theologizing has not taken seriously the fact that when we reflect theologically, we inevitably do so as embodied selves. Male theologians, in particular, have long assumed that the arena of theology is that of spirit and mind, far removed from the inferior, suspect body. Consequently, most theology has begun more deductively than inductively. It has begun with propositions and has attempted to move from the abstract to the concrete.

Indeed, for centuries it was not generally recognized that human bodies are active sources of meaning. Rather, it was believed that bodies were like cameras in a photographic process, simply recording external things mechanistically, things which were passed through the nervous system to form images in the brain according to physical laws. Now, however, there is reason to understand differently [14]. Our concern here is not primarily with "the body-object" as studied by the anatomist or physiologist but rather "the body-subject," the embodiment of our consciousness, our bodily sense of how we are in the world. Our concern is the interaction of the "givenness" of our fleshly realities and the ways in which we interpret them. It is our bodily sense of connections to the world, our bodily sense of the space and time we are in, our bodily knowing of the meanings of our relationships [22].

Body theology begins with the concrete. It does not begin with certain doctrinal formulations, nor with certain portions of a creed, nor with a "problem" in the tradition (though all of these sources may later contribute). Rather, it starts with the bodily experience of life – with making breakfast and making love, with beef burgundy and with bloated bellies of starving children, with daily calendars and with deadly carcinomas, with missives penned to those we love and with missiles aimed at those we fear. Body theology begins with the many big and little birthings and dyings we embodied beings encounter daily. Its task is critical reflection on our bodily experience as a fundamental realm of the experience of God. It is not, in the first instance, a theological description of bodily life from a supra-bodily vantage point (as if that were possible). Nor is it primarily a set of norms for the proper "use" of the body. Body theology necessarily begins with the concreteness of our bodily experience, even while it recognizes that this very concreteness is filtered through the interpretive web of meanings which we have come to attach to our bodily life.

We know the world through our embodiedness. As children, for example, we learn the most basic meanings of spatial relatedness through our sense of bodily location: prepositions such as "in," "with," "around," "under," "through," and "beyond" take on meaning through our sense of body placement. Moral knowledge is also bodily: if we cannot somehow viscerally feel the meanings of "justice" and "injustice," those terms remain abstract and unreal.

The way we feel about our embodiedness significantly conditions the way we feel about the world. Studies in body psychology, for example, disclose strong correlations between self-body connectedness and the capacity for ambiguity tolerance. Contrarily, there are also strong correlations between body alienation and the propensity toward dichotomous reality perceptions: the more one feels distant from the body, the greater the tendency to populate one's perceived world with sharply etched "either/or's" (either me or not-me, we or they, good or bad, black or white, communist or capitalist, heterosexual or homosexual) [6], [7], [8].[2] Our body realities do shape our moral perceptions in ways we have seldom realized.

"We do not just *have* bodies, we *are* bodies." This sentence is both a hopeful statement of faith and a lived experience. It is part of our faith heritage: Hebraic anthropology was remarkably unitary about the body-self, and when the Christian tradition is purged of its dualistic accretions its incarnationalism proclaims the unitary human being, too. At the same time the body-self unity is also an important part of our experience. We feel "most ourselves" when we experience such body-self integration. When, in illness, the body feels alien to us we say, "I'm not myself today." And we feel most fully ourselves when bodily connected with each other and the earth. The unitary body-self is not simply an abstract hope, a revelation "from outside" imposed upon a very different reality. We are able to articulate this faith claim and we are moved to do so precisely because this, too, is part of our body experience.

On the other hand, we do live between the times, knowing well the ravages of our body dualisms very personally but also socially and planetarily. We have been taught that not only is the body different from the real core of selfhood, it is also lower and must be controlled by that which is higher. Even our language is strongly dualistic: "I *have* a body" seems much more "natural" than "I *am* a body." And certain experiences – notably illness, aging, and death – seem to confirm the otherness of body. In those situations, the body seems radically different from the self. Though the body is "me," the body is also "it," a thing, a burden to be borne, sometimes an enemy. Then, though the body is "mine," I am also "its."

Thus, for good *and* for ill, the body has theological and ethical relevance in a host of ways. And our bodily experience is always sexual. Such experience is always sexual. Such experience obviously is not always genital – actually only infrequently so. But sexuality is far more than what we do with our genitals. It is our ways of being in the world as body-selves who are gendered biologically and socially, who have varying sexual orientations, who have the capacity for sensuousness, who have the need for intimacy, who have varied and often conflicting feelings about what it means to be bodied. It is all of this body experience that is foundational to our moral agency: our capacities for action and power, our abilities to tolerate ambiguity, our capacities for moral feeling. Our bodily experience lends considerable shape to our basic theological perspectives. The images and metaphors we find most meaningful to our God experience are inevitably connected to our life-long body experience.

If we are willing to accept the significance of the entire body-self in all of its sexual dimensions for doing our theology and ethics, we are still faced with the question, by what basic interpretive theory shall we approach this body life? Presently, the two major perspectives are commonly called social constructionism and essentialism. A social constructionist approach emphasizes our active roles as agents, influenced by culture, in structuring our bodily realities. It recognizes that the concepts and categories we use to describe and define our experience vary considerably in their meanings over time and among different cultures and sub-cultures. Further, it holds that the persistence of a particular interpretation of something depends not only on its correspondence to the reality being described, but at least as much on the usefulness of the concept, often its usefulness in social influence, power, and control [10].[3]

Symbolic interactionism, as an expression of social constructionism, emphasizes both the highly symbolic world in which we as human beings live and the relatively malleable nature of the realities we experience. Through language, symbols, and gestures we attach meanings to everyday acts and things. We do not respond to the things themselves so much as to the symbolic meanings they have for us. And these meanings are always social, arising, being modified and changed through social interaction [1].

Social constructionism can be contrasted with those approaches commonly labeled essentialist or empiricist, which stress the objectively-definable reality of topics of investigation. In such view, the body has its own intrinsic meanings. It has given nature and character, quite apart from what anyone believes about it. Sexuality, for example, may be acted out differently in different times and places, but there is something universal and constant about its core reality.

Both technical and popular discussions of sexuality still usually rest on essentialist assumptions. It is widely assumed that the sexual body is universally the same, always possessing certain sexual drives and needs. Those can be socially encouraged or thwarted in various ways and times, but they are fixed in their underlying essential nature. As with sexuality, so also with issues of bodily health and illness our cultural understandings have leaned strongly toward essentialism. The modern body has been heavily "medicalized." Using a biomedical model we give medical meanings to certain bodily conditions or behaviors, defining and classifying them in terms of health and disease. Then authorized medical practice becomes the primary vehicle for eliminating or controlling those conditions or practices defined as diseased or deviant. Such medicalization gives a privileged position to biological discourse and knowledge, assuming that disease and health are objective realities capable of universal definition and standardized practice.

In contrast, a growing minority approach contends that the modern understanding of sexuality has been socially constructed in particular historical contexts. In such interpretations, sexuality is not so much a "constant" – an essential human quality or inner drive – but rather a human potential for consciousness, behavior, and experience that can be developed and modified by social forces of definition, organization, and control. Thus, it makes sense to say that there are not simply variations of an underlying universal sexuality, there are indeed different sexualities [9].[4] Regarding health and illness, to be sure, there are given biological realities to most diseases, realities that social interpretations do not, of themselves, create. But what those diseases mean, how they are experienced, how they are treated – such things are never automatically given.

My own approach, while emphasizing social constructionism particularly as shaped by the symbolic interactionists, attempts also to recognize certain claims of the essentialists. Our sexual body-selves are subject to an enormous range of socially-constructed meanings that are extraordinarily plastic and malleable, and we need to understand those "realities" historically, contextually, and relationally. And, at the same time, there is still "something there" that is not *only* the creation of social discourse. It is true of our sexual orientations, for example. Contrary to some recent more extreme constructionist interpretations, I believe that homosexuality is not merely an artifact or construct of particular social structures at particular times and places. There is still something "given" about our sexual orientations, however significant the social meanings that shape their expression [12].[5] It is also true of our genders. That I have never menstruated, but that I do have penile erections does mean *something* for my interpretation of the world. Yet, just *what* these orientations and differently-sexed bodies mean is never fixed once and for all. And that is hopeful, for if sexual meanings are socially-constructed, they can also be reconstructed when they are not life-giving. Indeed, a holistic perception of the body-self has compelling reasons to hold both constructionist and essentialist perspectives together. Simply put, constructionism alone suggests spirit or mind without bodily reality, and essentialism alone suggests body without spirit.

An important parallel here is a relational value theory. Relationalism insists that value is neither simply objective (intrinsic to beings in and of themselves), nor is it simply subjective (the expression of our feelings about them). While there is some truth in both objectivism and subjectivism, it appears more adequate to say that value always arises out of and refers to relationships [20]. So also, with body ethics. Body values cannot simply be read off intrinsic meanings given in bodily life, meanings which are awaiting our discovery and application. Nor are body values simply social creations built on a completely plastic, malleable bodily reality. Rather, such meanings and values arise out of the interaction of bodily reality and our interpretive capacities as social, relational beings.

AN EXAMPLE: SEXUALITY IN MALE BODY THEOLOGY

Feminist scholarship reminds us that the literature of our western philosophical and religious traditions largely reflects the body and sexual meanings of male experience, though men have assumed those meanings to be generically human.

Currently we are trying to sort these things out. In that process it is crucial to ask questions about men's own body experience and their typical interpretations.

At the core of modern western masculinity, it appears, there are particular concepts of *reason* and *power* [21]. Regarding reason, two renowned philosophers of the Enlightenment, René Descartes and Immanuel Kant, are particularly revealing. Descartes was a thoroughgoing dualist: mind over matter, reason over body. The essence of human nature, he believed, is the reasoning mind, and the body is nothing more, nothing less than an intricate machine. (Note that the philosopher was saying more about a certain white Western historically-conditioned male experience than about human nature.) Descartes laid the philosophical foundations for modern medicine, and as a result medical science learned an enormous amount about the nuts and bolts of the body machine. When the body's carburetor went bad it could be taken out and replaced. But the result was clear: the body had nothing to do with the core of the self. That core was rational. And body feelings, body passions, body knowing were discounted.

Though Immanuel Kant's approach was different in many respects, there were similar results. For Kant the core of morality was reason and reason's control of the will. We are to make our acts correspond to universal reason, and in that process we learn to distrust emotions and feelings as bearers of authentic moral knowledge. The out-workings of Kantian reason are not difficult to see. If reason is to be in control, that means distancing from the body and its feelings. If reason is to be in control, in patriarchy that means males; women, children, and animals are insufficiently rational. The public is sharply divided from the private, for the public world is external and that of reason. The private world, the domain of women and children, is that of feeling and body. A man's home is his castle, but a man's world is the public arena. In all of this, men have become distanced from significant parts of their own lives and partially invisible to themselves.

Consider *power*. Men have traditionally seen power as some form of hierarchical control. "A real man" is a powerful man, and that means he is in control of his body, his feelings, his relationships, his woman, his children, his life. But such control limits much understanding. Understanding requires vulnerability and reciprocity. When I control my body, I develop little understanding of my emotional life, and then when my emotions really take hold, I am likely to be overcome by them. When I perceive myself as in control of others, I am unlikely to enter into their subjectivity; the controller knows little of the controlee's feelings and perceptions of the world, while the oppressed have wide eyes.

Power also has definite boundaries. It is limited in quantity. It is a "zero-sum" game: the more you get, the less I have. Hence, power assumes competition. And masculine in perception, power is shaped by individual boundaries more than by relationships. Ronald Reagan, the cowboy-actor, as president effectively both symbolized and appealed to such understandings of individualism and male power. Part of his effectiveness lay in his apparent triumph over his own body. He looked youthful in spite of his years, could joke even after being shot, could turn his failing hearing into an asset, and could chop wood and ride horses when his contemporaries took to their rocking chairs.

But when I am alienated from my bodily feelings there is a terrible toll. When I am detached from my own bodily reality I am prone to abstractions about the bodily reality of others as well as myself. In a still male-dominated medical profession dying patients frequently are overtreated; their embodied realities have faded into charts, diagnoses, and medico-moral abstractions. So also, in war-time, military language speaks with disembodied abstractions. Enemies lose their concrete humanity and become "gooks"; soldiers talk of destroying villages in order to save them.

We who are men have learned some profound ambivalence about our own embodiedness. As a corollary, we live with a lot of mixed feelings about our sexuality. Consider, for example:

- We are proud of our sexuality; it is a symbol of masculine power. And we genitalize much sexual meaning so that "he's got balls" means manhood. But we are also anxious about our sexuality. The thought of impotence can paralyze us. And the perception of women as more sexual than we, and, for heterosexual males, the perception of gays and lesbians as more sexual than we, both frightens and angers us.
- We are confused about intimacy. We tend to equate intimacy with "having sex." But real intimacy is problematic because early we learned the lessons of masculinity by breaking the erotic bonding with our mothers, by distinguishing ourselves over against girls, and by keeping emotional distance from other males so we wouldn't seem "queer." Ever since then, emotional closeness with anyone can threaten our tenuous sense of masculinity.
- We believe that we love sex, we love body pleasure, we can never get enough of these things, and all of that is part of being a man. Yet the sex we learned in boyhood was not a way of nourishing and pleasuring the body-self, but a release from a drive, with such releases to be measured by the statistics of performance and competition.

These, I think, are a few of the fairly common sexual-body experiences and meanings of men in our culture. They are neither universally human experiences nor universal male meanings. The meanings are based on experiences of having male bodies, but they are particular masculinity images that are historically and culturally relative. And they are relative regarding the differing contexts of sexual orientation, race, and social class in which these cultural masculinities are experienced. What we are about is the task of unmasking our male experience, because we who are men have become largely invisible to ourselves.

Serious body theology is long overdue for men. It must begin with the concreteness of our body experiences and meanings. But it cannot stop there. It must take that particular body stuff and reflect on it all in light of our relatedness to God, each other, and the earth. In patriarchy, taking "man" as generic human has meant that we have seldom explored and little understood our distinctively male experience. Hence, we have been uncritical. And that leads to enormous problems in the practice of sexual wholeness and justice.

That body theology is needed in our time seems apparent from the earlier discussion of the seven deadly sins. Recall how the power of each of them depends in no small measure on body anxiety and fear. Useful theology must do more than pronounce, "The body is good and one with the spirit; be not afraid!" Useful theology must move deeply into our body experiences themselves and interpret them in dialogue with the experience of God's reality.

But body theology must also deal with gender experience. Consider how the seven deadly sins cannot be understood without grasping the dominant role of male meanings in them. And consider how the seven virtues cannot be released more fully in their authentic power without critical reformation of male body meanings. For further illustration of our need to unmask the male experience theologically, consider the following possibilities.

- Because we have capitulated to a one-sided phallic masculinism in our theology, we have had centuries of imaging God as unilateral, non-relational power, glorified by the weakness and dependency of humanity. And countless people have found such Christian spirituality lacking, deficient in the power of mutuality.
- Because we have attributed to God a one-sided power, we have compensated with a one-sided and unilateral definition of Christian love: sheer agape, sheer self-giving. In the process, we have largely lost the goodness of the erotic, we have confused self-love with selfishness, and we have impoverished our humanness.
- Because we have absorbed so much spiritualistic dualism, we have tried to minister to persons as though they were disembodied spirits. And countless people have found their pressing body and gender issues ignored by the Church.
- Because we have ignored the marvellous completeness of each of our bodyselves in the image of God, we have adopted theologies that insist upon male-female complementarity. And these theologies have particularly oppressed women, single persons, gays, and lesbians. (Even Jesus, mind you, was not whole by such definition.)
- Because we have capitulated to gender stereotypes, and particularly have allowed males to embrace an oppressive phallicism, we have also capitulated to an alarming, tragic teenage pregnancy rate by teaching boys to prove their manhood without responsibility.
- Because we have allowed phallic images to dominate our perceptions of reality, we have created a rape culture where every woman every day of her life must be aware of the threat of male violence; and we have undergirded a culture of social violence and environmental exploitation, to the danger of life on a fragile planet.

BODY REVELATION

If body and sexual meanings are in large measure socially-constructed, they can be reconstructed. Flesh is necessary medium of Word, and body is medium of revelation. Consider, for example, how our bodily realities might be revelatory experiences of the divine *shalom,* God's inclusive community of right-relatedness [11]. Each of us is composed of more than a trillion individual cells, all trying to work together and maintain one another. Our bodies are communities – with their own ventilation systems, sewage systems, communications networks, heating unions, and a billion miles of interconnecting streets and highways.

Our bodies are not only communities in themselves, but also communities in relationship with the earth. Our body fluids have the same chemicals as the primeval seas – we carry those seas within us. Our bones contain the same carbon as the rock of the oldest mountains. Our blood has the sugar that once flowed in the sap of now-fossilized trees. The nitrogen binding our bones together is the same as that which binds nitrates to the soil. Our bodies, if we listen and feel, tell us that we are one with creation. And, even when cancer eats at our vitals and when genetic disease deforms infants, there is still revelation in the midst of tragedy. The tragic is tragic precisely because it takes its meaning from a more original vision of harmony and interconnectedness.

Teilhard de Chardin once commented that we have been taught to understand our bodies as fragments of the universe, as pieces completely detached from the rest, handed over to us to inhabit. But we must learn, he said, that the body is the very universality of things. My body is not part of the universe that I possess totally – it is the totality of the universe that I possess partially [23].

With confidence that body experience can be the medium of revelation, how might we move into body theology? How might we move into the dialectic between body revelation and revelation as understood in the religious community? Obviously, there is no single path. Let me suggest just one, and here I must speak from my own particularity as part of the Christian tradition, committed to trying to understand and experience more fully the reality of incarnation.

Incarnation in Webster's primary definition simply means embodiment – being made flesh. Religiously, it means God's embodiment. Christianly, it means "Christ." In particular, it means Jesus as the Christ, the expected and anointed one. Through the lens of this paradigmatic embodiment of God, however, Christians can see other incarnations: the christic reality of other human beings in their God-bearing relatedness. Indeed, the central purpose of christology, I take it, is not affirmations about Jesus as the Christ. Rather, affirmations about Jesus are in the service of revealing God's christic presence and activity in the world now.

This is a more kenotic approach to christology than has often been assumed. Indeed, traditional christologies frequently have raised difficult problems. The formula of a hypostatic union of two natures was largely based upon and has perpetuated a dualistic metaphysic. Further, confining divine incarnation exclusively to Jesus made of him a docetic exception to humanity and disconnected Christ from our experience. Focusing upon *who* Jesus was has relegated his actions and relationships to secondary importance. Centering salvation history in Jesus as the only Christ has buttressed Christian triumphalism and oppressed many persons [5].

In short, traditional christology largely succumbed to a one-sided essentialism. It assumed that, through unilateral divine decision and action, there was the intrinsic essence of God in Jesus of Nazareth, quite independent of Jesus's own experience, contexts, relationships, and his own and others' interpretations.

I understand Jesus the Christ to be a confessional interpretation and claim of Christians. This one is *our* paradigmatic experience of the christic presence. Here we have encountered the one who particularly embodied and thus still clarifies and releases God's presence, action, and relatedness. But he serves to clarify and release all of that in ways that nurture many other christic embodiments. The union of God and humanity in Jesus was a moral and personal union – a continuing possibility for all persons. Incarnation is always a miracle of grace, but the essence of miracle is not "interference" in the "natural" world by the "supernatural." It is the gracious (hence miraculous) discovery of who we really are, the communion of divine and human life in flesh. And one of the central criteria of christological adequacy must be the moral test: does this interpretation of what christology is all about and what the Christ means result in bodying forth more of God's reality now? Does it create more justice and peace and joyous fulfillment of creaturely bodily life?

We who are men particularly need a gracious resurrection of our bodies. It is our hope for the postpatriarchal male. Resurrection of the body now, insofar as we experience it, is the gracious gift of a fundamental trust in the present bodily reality of God, the Word made flesh now in our bodily relatedness. Resurrection is the movement from dualistic alienation to incarnation. Incarnation is the meeting of the Creator and creature in and through matter. In human beings, the christic reality is the communion of the divine and human loving in and through the living body-self. To embody (to body forth) God's agapaic, philial, erotic love in the world – the love that is expressed in justice and peace and connectedness – then, is both our possibility and our moral imperative. And with the resurrection of the body comes the experience of divine immediacy and immanence, God's continuing incarnation.

All of this suggests that the human body is language and a fundamental means of communication. We do not just use words. We are words. This conviction underlies Christian incarnationalism. In Jesus Christ God was present in a human being not for the first and only time but in a radical way that has created a new definition of who we are. In Christ we are redefined as body words of love. Such body life in us is the radical sign of God's love for the world and of God's immediacy in the world [4], [24].

This incarnational perspective, only briefly sketched, is one way of beginning to move into the questions of the ultimate meanings of our body and sexual experience. There are other ways. But whatever paths we follow, we will find ourselves not simply making religious pronouncements about our sexuality and body lives. We will enter theologically more deeply into this experience, letting it speak of God to us, and of us to God.

The significance of all this has not escaped Toni Morrison in her Pulitzer Prize novel *Beloved.* A central character is Baby Suggs, grandmother and holy woman of the extended family who had escaped from slavery in the South. Speaking to her people, Baby Suggs "told them that the only grace they could have was the grace they could imagine. That if they could not see it, they would not have it. 'Here,' she said, 'in this place, we flesh; flesh that weeps, laughs; flesh that dances on bare feet in grass. Love it. Love it hard. Yonder they do not love your flesh. They despise it .... You got to love it. You.'" ([16], p. 88). Note carefully Baby Suggs's counsel. The only grace you can have is the grace you can imagine. If you cannot see it, you will not have it. That invites imaginative body theology.

United Theological Seminary of the Twin Cities
New Brighton, Minnesota, U.S.A.

NOTES

[1] In this section I repeat in abbreviated and altered form some of the points I have made in [17].

[2] Fisher says, "Although we are still in the early stages of understanding body image phenomena, we have discovered that body attitudes are woven into practically every aspect of behavior. The full range of their involvement cannot be overstated" ([8], p. 626). There is reason to assume that our body-self experiences as females and as males are more similar than dissimilar. Here I am dealing in the basic human commonalities of body experience. Nevertheless, there are also certain important differences of world perception related to our different sexual biologies as well as to our sex-role conditioning. I have described some of these with their possible theological-ethical implications in [19].

[3] Social constructionism is indebted to a number of different intellectual schools such as symbolic interactionism, literary deconstructionism, symbolic anthropology, existentialism, phenomenology, and, in general to social psychological theory. It may be contrasted with empiricism and positivism which stress the objectively definable reality of topics of investigation.

[4] While others (e.g. P.J. Naus, J. Weeks, K. Plummer, W. Simon, and J. Gagnon) have also pursued a largely social constructionist or symbolic interactionist position, arguing against the notion that the body is universal in its sexual drives and expressions, this still remains a minority position among sexuality scholars. For my own application of this approach to sexuality see [17]. Social constructionist feminists are well represented by Carter Heyward [13].

[5] Greenberg gives too little credence to the givenness of sexual orientation early in life. A good critique of Greenberg is found in J. Boswell [2]. While in this essay I am generally accenting the social constructionist side, I believe it also important to recognize that there are, indeed, certain more objective bodily realities underlying our interpretations.

BIBLIOGRAPHY

1. Blumer, H.: 1969, *Symbolic Interactionism: Perspective and Method,* Prentice-Hall, Englewood Cliffs, NJ.
2. Boswell, J.: 1989, 'Gay History', *The Atlantic,* 263/2.
3. Brown, J.C. and Bohn, C.R. (eds.): 1989, *Christianity, Patriarchy and Abuse: A Feminist Critique,* Pilgrim Press, New York.
4. Davis, C.: 1976, *Body as Spirit: The Nature of Religious Feeling,* Seabury, New York.
5. Driver, T.F.: 1982, *Christ in a Changing World,* Crossroad, New York.
6. Fisher, S.: 1970, *Body Experience in Fantasy and Behavior,* Appleton-Century-Crofts, New York.
7. Fisher, S.: 1973, *Body Consciousness,* Prentice-Hall, Englewood Cliffs, NJ.
8. Fisher, S.: 1970, 1986, *Development and Structure of the Body Image,* Vol. 2, Lawrence Erlbaum Associates, Hillsdale, NJ.
9. Foucault, M.: 1978, *The History of Sexuality,* Vol. I: An Introduction, Pantheon, New York.
10. Gergen, K.J.: 1985, 'The Social Constructionism Movement in Modern Psychology', *American Psychologist* 40, 266–275.
11. Gray, E.D.: 1979, 1981, *Green Paradise Lost,* Roundtable Press, Wellesley, MA.
12. Greenberg, D.F.: 1989, *The Construction of Homosexuality,* University of Chicago, Chicago, IL.
13. Heyward, Carter: 1989, *Touching Our Strength: The Erotic as Power and the Love of God,* Harper & Row, New York.
14. Johnson, D.: 1983, *Body,* Beacon Press, Boston, MA.
15. Kazantzakis, N.: 1965, *Report to Greco,* P.A. Bien (trans.), Bruno Cassirer, Oxford.
16. Morrison, T.: 1987, *Beloved,* New American Library, NY.
17. Nelson, J.B.: 1978, *Embodiment: An Approach to Sexuality and Christian Theology,* Augsburg, Minneapolis, MN.
18. Nelson, J.B.: 1984, 'Religious Dimensions of Sexual Health', in J.M. Joffe et al. (eds.), *Readings in Primary Prevention of Psychopathology: Basic Concepts,* University Press of New England, Hanover, NH & London.
19. Nelson, J.B.: 1988, *The Intimate Connection: Male Sexuality, Masculine Spirituality,* Westminster, Philadelphia, PA.
20. Niebuhr, H.R.: 1960, *Radical Monotheism and Western Culture,* Harper & Brothers, New York.
21. Seidler, V.J.: 1989, *Rediscovering Masculinity: Reason, Language and Sexuality,* Routledge, London.
22. Shapiro, K.J.: 1985, *Bodily Reflective Modes: A Phenomenological Method for Psychology,* Duke University Press, Durham, NC.
23. Teilhard de Chardin, P.: 1968, *Science and Christ,* Harper & Row, New York.
24. Vogel, A.A.: 1973, *Body Theology,* Harper & Row, New York.

BARBARA H. ANDOLSEN

WHOSE SEXUALITY? WHOSE TRADITION? WOMEN, EXPERIENCE, AND ROMAN CATHOLIC SEXUAL ETHICS

This essay focuses on the position of women vis-à-vis "tradition" – the process by which a religious community hands on, from generation to generation, the wisdom of the past, in this case, wisdom concerning human sexuality. From my perspective as a woman, the materials concerning sexuality usually encompassed by the phrase "the Roman Catholic tradition" are a painful and sometimes repulsive collection. Misogyny, distrust of the body, and antipathy toward sexuality are found in too many classic Catholic theological sources. Reexamining the tradition from a perspective that affirms women's struggles for sexual wholeness, I have gained a heightened sense of the scope, complexity, and ambiguity of my church's moral memories.

Throughout this essay, I will draw upon examples from the tradition of my religious community, Roman Catholicism. However, questions about tradition raise critical issues for other religious groups, too. Tradition is particularly important as a basis for ethical reflection about sexuality in Judaism, the Eastern Orthodox churches, and Roman Catholicism. It also exerts a more subtle influence on Protestant ethics, despite the propensity of some Protestant ethicists to appeal primarily to a narrow range of historical strata among their traditions, i.e., the Biblical traditions. In all these religious groups, women have rarely possessed the social and religious power necessary to be effective participants in the shaping of corporate religious memories. What Judith Plaskow says of her religious tradition is true, in a somewhat different fashion, for Christianity, too.

> The need for a feminist Judaism begins with hearing silence. It begins with noting the absence of women's history and experiences as shaping forces in the Jewish tradition .... Women have lived Jewish history and carried its burdens, but women's perceptions and questions have not given form to scripture, shaped the direction of Jewish law, or found expression in liturgy ([38], p. 1).

Women's perspectives and questions have not contributed in a substantial way to the formation of Christian tradition, either.

In this essay, I approach the examination of Catholic tradition as a feminist. I insist that women must become active participants in the formation of the tradition for their own sake as members of the religious community and for the good of the community as a whole. An adequate assessment of Roman Catholic sexual ethics must include explicit attention to the sexual well-being of women. A renewed sexual ethic must speak meaningfully to Catholic women in very diverse circumstances. Among (but not exhaustive of) the differences which must be taken into account are marital status, fertility or infertility, sexual

Ronald M. Green (ed.), Religion and Sexual Health, 55–77.

preference, race, national origin, and economic class. I have tried to remain aware of these variations in writing this essay, but ultimately defining a new Catholic ethic will require a dialogue among many different women.

One feminist criterion for evaluating the church's sexual ethic is a requirement that the church's teachings must acknowledge and enhance the moral agency of women. Basic to all moral agency is what Carol Robb has termed "the capacity for responsible self-direction" ([23], p. xvi). Women and men need an ethic that calls them equally to responsible self-direction in sexual matters. In this essay I propose a substantially revised Roman Catholic sexual ethic that places responsibility for morally appropriate sexual restraint on each individual as a moral agent responsible for his or her sexual behavior and accountable to others for the consequences of his or her sexual choices.

By contrast, throughout most of Catholic history, the official teachers of the church have perpetuated an ethic of male control over female sexuality. The "fathers of the church" have sought to constrain women's sexuality for the sake of men's spiritual well-being. The aid of Catholic women was enlisted in protecting men from their own sexual urges. Female purity and chastity safeguarded men's virtue as well as women's own. The final vestiges of such control of female sexuality are found in the fading, but still perceptible, traces of the "double standard" – male sexual lapses are an understandable result of men's inability to control their raging sexual urges, while women's falls from purity represent grave female failures to contain not only their own sexual passions, but those of their male partners as well.[1]

In addition to proposing an ethic of responsible self-direction for women and men, I stress a code of sexual ethics consistent with a recognition of the equal moral worth of women and men. Exponents of the church's official sexual ethic would insist that their sexual norms are uniquely consistent with the human dignity of man and woman properly understood. Proponents of the hierarchy's teachings contend that they offer a reliable code of sexual morality rooted in the objective natural law. They claim that their moral teachings are congruent with the basic, unchanging structures of human sexuality.[2]

The natural law tradition in the Roman Catholic church has been articulated largely from the vantage point of a clerical caste in the church, because, at least until recently, access to theological education has been closely linked to the clerical state of life. Thus, most of the persons who possessed the formal credentials necessary to contribute to expositions of sexual ethics were celibate men who viewed sexual issues from a limited range of male perspectives. What those in sympathy with the official teachings of the church regard as an objective, rational analysis of the trans-historical, natural (moral) law concerning human sexuality is, in reality, a series of culture-bound explications of those sexual norms which seemed compelling to certain members of the celibate clergy.

The reorientation of tradition that I call for here requires increased attention to the experience of *ordinary folk, especially women.* Contemporary Roman Catholic theology emphasizes that all believers, with their multifaceted experiences of human sexuality, *are the church.* Therefore, the entire community of believers should be at the center of the Roman Catholic tradition. There are differing responsibilities within the church for public articulation of tradition. Nevertheless, it is the witness of Catholic peoples – women and men, gay and straight, sexually active and celibate – to the human wholeness which God holds out to them that provides the central threads in Catholic tradition.

At present, we are far from having adequate interpretations of the living tradition of the entire church. It is very difficult, given presently available historical resources, to recover and reinterpret the faith experiences of those generations of women and men who have been the ordinary members of the church throughout the centuries. So, far too often, Catholic ethicists reduce tradition to a litany of carefully selected excerpts from the writings of prominent theologians and from the statements of popes and church councils. Ethicists who make the standard appeals to tradition become entrapped in a male-defined, elitist intellectual heritage. There is a profound irony in this intellectual captivity. For there is significant hermeneutical evidence that the Jesus community, which was the genesis of the entire Christian faith, was not elitist, at least arguably not patriarchal, and not preoccupied with control of sexual behavior [16], [10].

My review of the "official" tradition of the Roman Catholic church on sexuality in preparation for this essay was profoundly alienating. It often meant seeing human sexual behavior through the eyes of men who are distrustful of women. It frequently involved the disorienting experience of viewing persons like me – others with female bodies – as *the Other.* The images of women offered by the "fathers of the church" seemed as distorted as the reflections in a fun house mirror.

TERTULLIAN'S MISOGYNY REEXAMINED

Christian men, especially celibate men, have frequently dealt with ambivalence about their own erotic feelings by projecting sexual passion onto women whom they stereotyped as wanton and seductive. There is little that could contribute to female wholeness in those powerful streams of religious tradition in which men view women as seductive bodies that represent serious obstacles in the male path to salvation. Yet, even the most misogynist theological statements may be more complex than they first appear. Paradigmatic for me of the woman-as-evil-temptress view are statements from Tertullian's essay concerning women's appearance – their cosmetics and dress – *De Cultu Feminarum.*

Tertullian was a second century (married) theologian who stressed modesty and chastity as crucial virtues necessary for both men and women to preserve holiness in the interval between baptism and death in the grace of God. Since Tertullian viewed rigorous sexual self-restraint as essential to the Christian life, he strongly discouraged Christian women from engaging in practices that heightened sexual attractiveness. He decried women's use of cosmetics and alluring clothes. In one of his more temperate passage, he advised that a Christian woman should achieve a balance in her appearance – that she should strive for "natural and demure neatness" (II, v, i).

However, in this essay he went beyond a denunciation of fashions that exaggerated a woman's sexual appeal; he recommended that Christian women disguise even *naturally attractive* bodily features. Love of the male neighbor entailed for Tertullian that a woman disguise her physical beauty, so as not to arouse his lust. Tertullian advised: "you must not only shun the display of false and studied beauty, but also remove all traces of natural grace by concealment and negligence, as equally dangerous to the glances of another's eyes." It is not altogether clear what Tertullian would have a naturally beautiful Christian woman do, because he added that he saw no virtue in bodily filth and he was not advocating "an utterly uncultivated and unkempt appearance" (II, ii, v).

In yet another section of this same essay – a section characterized by particularly strong rhetoric – Tertullian descended to misogynist excess. He suggested that new female converts, rather than continuing to practice customs of fashion that accentuated their physical attractions, should have adopted the austere garb of one in mourning – mourning for her sins. Converts might have been expected, he claims, to neglect their appearance, "acting the part of mourning and repentant Eve in order to expiate more fully by all sorts of penitential garb that which woman derives from Eve – the ignominy, I mean, of original sin and the odium of being the cause of the fall of the human race" (I, i, i). Women's natural allures are connected with the sexual appeal of the first woman, Eve, who, in Tertullian's mind, caused the downfall of the human race.

Determining the import of these passages from the tradition for contemporary women in a fair way is a complex challenge. This document is addressed to upper-class, second-century, Christian women, especially to female converts who must learn wholly to reorient their lives in a manner consistent with their new religious commitment. Therefore, *De Cultu* emphasizes issues of women's appearance and virtue. Throughout his writings, Tertullian advocates a high standard of sexual self-restraint for men as well as women. In other works, Tertullian discusses Adam's pivotal role in original sin or describes how humanity – male and female – are heirs to the fall [8]. Moreover, his views on sexual activity reflect a particular ancient Christian mentality concerning sexual indulgence and sexual restraint. This ancient Christian view is in some ways intractably foreign to modern experience ([4], p. xv). In addition, Tertullian's penchant for excessive rigorism ultimately led him to a group called the Montanists who were rejected as heretics by the main body of Christians.

Still, Tertullian is representative of a major strand in Catholic tradition insofar as he places a special burden on women to take responsibility for male sexual behavior without any reciprocal demand addressed to men. He does ask men to curb their sexual passion, even during intercourse with their wives, and to consider a higher life of sexual continence. He also admonishes men to be less vain about their appearance. Although Tertullian describes male vanity more briefly as antithetical to the sober virtue that should characterize the lives of Christian men (*De cultu feminarum,* II, viii), he does not dwell on male sexual allure as a danger to *women's* souls. Here, Tertullian stands near the source of one unfortunate stream in Catholic sexual tradition. This is the stream of tradition in which women are viewed as seductive sexual objects whose sexuality must be constrained for the sake of men's spiritual well-being.

What Judith Plaskow has said of Judaism is true of Roman Catholicism as well, albeit in a somewhat different fashion: "Men define their own sexuality ambivalently – but they define it. And men also define the sexuality of women which they would circumscribe to fit the shape of their own fears, and desire for possession. Women must carve out a sense of sexual self in the context of a system that – here [on issues of sexuality] most centrally – projects them as Other, denying their right to autonomous self-understanding or action" ([38], pp. 191–92).

BROADENING THE TRADITION

The tradition, as it has usually been framed by Catholic ethicists, almost never includes the self-understanding and practices of autonomous women. Instead, as I have indicated, the history of Roman Catholic ethics, including the history of teachings on sexuality, remains largely a chronological account of the ideas of prominent theologians. Throughout much of the church's history very few women have been allowed access to those forms of education that were a prerequisite to making one's view of sexuality intelligible in theological language. The insights of women have rarely found a place among those documentary memories which have been preserved for future generations.

The history of Roman Catholic ethics, in particular, and Christian ethics, in general, is in much the same state as the larger academic discipline of history was several decades ago. Then, secular history was dominated by the study of the activities of diplomats, military leaders, and leaders of state who were overwhelmingly men. The history of Christian ethics *remains* primarily the history of (male) theologians. In Roman Catholic ethics, the statements of ecumenical councils and popes receive heavy emphasis. Secular history as a discipline has been challenged and transformed by new work which deals with social and family history. Many secular historians now give careful attention to the experiences of women and less powerful men. Still, few ethicists interested in the history of Christian ethics have adopted approaches similar to those developed by these historians.

Catholic ethicists should become involved in collaborative scholarship that integrates the experiences of women and less powerful men throughout the history of ethics. In preparation for such a long-term, cooperative scholarly enterprise, I would like to pursue briefly two examples of how our understanding of the Roman Catholic ethical tradition might be broadened by attention to the experiences of these other members of the church. I will explore, first, the bodily piety of certain late medieval women and, second, the protests about the imposition of clerical celibacy from the fourth through the twelfth centuries.

THE BODY AND MEDIEVAL WOMEN'S PIETY

During the later middle ages, women were drawn to the Eucharist as an element in a spirituality that paid special attention to and valued physicality. As with other sacraments of the Roman Catholic church, the Eucharist is rooted in sensual, bodily experience. The doctrine of transubstantiation affirms that Jesus Christ is truly present body and blood in the bread and wine of the Eucharist. This doctrine establishes "an intimate interconnectedness of the physical and spiritual by which physical food can nourish spiritual life ... the boundaries of body and soul are not absolute, but permeable" ([31], p. 107). Ethicist Giles Milhaven notes: "Unprecedented in Europe was the eagerness of the [medieval] faithful, especially women, to receive the Eucharist, to eat of the bread, drink of the wine, and thus receive Christ within them" ([32], p. 346). Women pressed for more opportunities to be united with the body of Christ through more frequent reception of the Eucharist. Some women even rushed from church to church while Masses were being said in order to receive Communion as often as possible within the same day.

An intense desire to be united with Christ through receiving and becoming one with his body in the Eucharist was consonant with a powerful erotic spirituality in which Christ was received as a lover.[3] The medieval mystic Hadewijch gave particularly eloquent expression to this spirituality when she described one unusually memorable experience of her union with Christ through reception of the Eucharist. Present at morning service on Pentecost, Hadewijch was in a state of keen anticipation of union with Christ. "My heart and my veins and all my limbs trembled and quivered with eager desire ...." She longed "to have full fruition of my Beloved, and to understand and taste him to the full." After she had received Communion, Christ, "came himself to me, took me entirely in his arms, and pressed me to him, and all my members felt his [body] in full felicity, in accordance with the desire of my heart and my humanity. So I was outwardly satisfied and fully transported." Then it seemed to her that she and Christ melted into one. "After that I remained in a passing away in my Beloved, so that I wholly melted away in him ...." ([24], pp. 280–82). Certain medieval women unabashedly connected physicality and eroticism with the knowledge of God. Milhaven suggests that this bodily piety of medieval women was not incorporated into the high theological tradition of their time or of later ages. "Thomas and other theologians ignored the wisdom growing from sexual love and pleasure" ([32], p. 362).

Yet, to correct the exclusion of medieval women's perspectives from the tradition is again to uncover a complex history. The late medieval female spirituality did include a special attention to physicality – both women's experiences as bodily selves and Christ's bodily experiences. Particularly through reception of the Eucharist, medieval women mystics identified themselves with a Christ who fed the whole community, especially the poor and suffering, with his body. However, that same spiritual tradition was also marked by a propensity to extreme bodily asceticism. As historian Caroline Bynum states, "women's devotion was more characterized by penitential asceticism, particularly self-inflicted suffering" ([5], p. 26). Some of these women deprived themselves of sleep, starved themselves, flogged themselves, and deliberately consumed lice and pus from the wounds of seriously ill patients for whom they cared. Indeed, women were disproportionately represented among those saints who died as a result of severe ascetic practices ([5], p. 76). Bynum emphasizes that such practices can only be understood properly within their historical and cultural context: "... what modern eyes see as self-punishment ... [was to medieval people] imitatio Christi – a fusion with Christ's agony on the cross" ([5], pp. 211–12). She also reminds the reader that medieval people could frequently do little to ameliorate human suffering ([5], p. 245). Therefore, the ability to give suffering spiritual meaning was a precious capacity. The complex interrelationship of physicality, sensuality, concern for the poor and self-inflicted suffering in the religiosity of certain medieval women poses a challenge to me as a twentieth-century ethicist who wants to affirm the moral worth of bodily well-being and pleasure.

PRIESTLY CELIBACY AND THE PURPOSES OF MARRIAGE

The prolonged protest against the imposition of mandatory celibacy for the clergy is another interesting strand in the history of Catholic sexual ethics. Ethicists can enrich their understanding of tradition by using a technique of feminist historiography, i.e., by viewing history from the underside – from the perspective of the devalued members of the community. Ethicists should search for moral wisdom known to historical "losers."[4] In this case, I have in mind the wisdom of those who, in the Roman Catholic tradition at least, struggled unsuccessfully to preserve a married clergy.

The imposition of celibacy involved a protracted struggle, which culminated in the eleventh and twelfth centuries. There were many episodes of resistance. In the fifth century, for example, Synesius of Cyrene agreed to become a bishop only on the condition that it be clearly understood that he would continue to have an active sexual relationship with his wife. He refused to continue their relationship "surreptitiously like an adulterer." He expressed the hope that he would father "many virtuous children" ([17], p. 199). Thus, this upper-class church leader placed a high value on marriage and procreation.

Even a bit earlier, by the fourth century, there was a fierce intellectual controversy over the relative merits of virginity and marriage. Unfortunately the defense of marital sexuality offered by Helvidius, Vigilantius and Jovinian are known to us only second hand through the vituperative rejoinders of Jerome. This well-known case of a loss of primary historical source materials highlights the difficulty of reconstructing the full tradition of the Catholic church. We are far from having sound documentary records from all the persons and about all topics of interest to modern scholars.

After studying eleventh and twelfth century materials from the continuing debate concerning celibacy, historian Anne Barstow indicates that "married clergy and their supporters left a valuable legacy of arguments affirming the goodness of ... human sexuality, affirmations seldom enough found in medieval theological writings" ([2], p. 155). For a twentieth-century ethicist like me to consider the protestors as a positive part of the tradition, involves immersing myself in the ambiguities of history. Those who opposed mandatory clerical celibacy affirmed positive aspects to sexual activity within marriage and insisted that sexual intimacy with a woman did not prevent a man from functioning effectively as a religious leader.

Yet, medieval arguments against imposition of celibacy present strange and unappealing elements to a modern interpreter as well. A defense of marriage was not necessarily an affirmation of the inherent goodness of sexual pleasure. Reviewing notions of (marital) sexual relationships in Christianity, historian Margaret Miles notes: "A dichotomy between marriage and pleasure was assumed in many writings – marriage being good and pleasure bad. In marriage, sex was tolerated largely for purposes of the continuity of the human race, although some authors can imagine a companionship between husband and wife that is a good in itself." She continues: "Clearly, it was never sexual intimacy or activity itself that Christian authors valued and were grateful for" ([31], p. 155).

In the middle ages, married clergy and their supporters did not appeal to an ideal of marriage as a realm of specially valuable interpersonal emotional and sexual intimacy. Instead, they defended marriage as an institution that let people direct their fierce sexual urges into a socially acceptable channel, thus avoiding sexual sins. The protestors against celibacy frankly acknowledged that marriage was good as a socially necessary relationship that enabled persons to reproduce their kind in a responsible way. They pointed out that for the (nonmonastic) lower clergy, marriage could be an essential economic arrangement in which a man and a woman pooled gender-specific productive skills that were crucial for the day-to-day survival of a household.

The pragmatic concerns of the protestors point toward another area of family and sexual history which ethicists need to reexamine. Even "love" is not a timeless norm in Christian sexual ethics; it has specific meanings in particular historical contexts. A widespread cultural view that loving affection and interpersonal intimacy are primary aims of marriage represents a specific development in modern European and American history.[5] The social ideal of marriage as a loving partnership with sexual intercourse serving an emotionally and spiritually unitive function arose under particular historical circumstances. It was rooted in two historical developments, among others. First was the Reformation and later Protestant emphasis on marriage as a vocation for almost all adults with an increasing emphasis on the "spiritual intimacy" between husband and wife ([42], p. 101). Second was the privatization and sentimentalization of the home that occurred as many types of economic production were removed from the household through industrialization. In the newly privatized home of the middle class, the emotional bonds between family members took on novel importance and intensity. As ethicist James Nelson has recognized:

> The notion of romantic marriage is part of a larger social revolution in the West which, beginning in the seventeenth century, also led to the development of the capitalistic economy and the democratic state. It was a basic reordering of the traditional ways in which persons were understood to relate to each other. Before that time, the more usual assumption had been that people must adapt to society. With that revolution, however, it became commonly believed that social institutions could and should be changed to meet [individual] human needs. That applied to marriage and the sexual relationship ([35], p. 108).

Christian ethicists need to probe the implications of this social revolution in marriage and the family far more carefully than we have yet done. I cannot provide an exhaustive treatment of that subject here, but I would like to call attention to several implications.

It appears that this change in the understanding of spousal relationships began as a modern middle-class experience, articulated in the sermons and writings of Protestant ministers, among other sources. In the later stages of this social transformation of the purposes of marriage, medical experts and a newly emerging class of psychological experts played an increasingly authoritative role. By the late nineteenth and early twentieth century in the United States, the social influence which religious authorities exerted over generally accepted societal norms for sexual behavior had diminished. Medical experts took on a correspondingly greater degree of social power to shape cultural notions of acceptable sexual behavior.[6] The message that married couples received from certain medical authorities was an uplifting one. "By the late Victorian period, medical writers increasingly spoke of sexuality as an important means for enhancing the spiritual unity of husband and wife" ([12], p. 69). The medical advice literature "called attention to the importance of sexuality in personal life, often elevating it as a powerful force imbued with possibilities for heightened marital intimacy and even spiritual transcendence" ([12], p. 67).

It is this modern, romantic ideal of sexual intercourse as an expression of total interpersonal unity that was belatedly adopted in twentieth-century, magisterial Roman Catholic ethics.[7] An indication of the incipient impact of modern views of marriage and sexuality on the teachings of the magisterium is found in the 1930 encyclical, *Casti Connubïi*. Having already reiterated the long standing view that the procreation and education of children was the primary end of marriage, Pius XI went on to praise "the love of husband and wife which pervades all of the duties of married life and holds pride of place in Christian marriage" (Paragraph 23; see also Paragraphs 12–17). Such Christian love between spouses is not based on "the passing lust of the moment." Rather such love represents "a spiritual unity – aimed at the perfection of the character of spouses" (Paragraph 23). Pius even asserted that a marital love that uplifts the moral and spiritual character of the spouses "can in a very real sense ... be said to be the chief reason and purpose of matrimony" (Paragraph 24). Yet, there remains a serious tension between an incipient acceptance of this modern view of spousal love and an older Catholic view that procreation is the primary end of marital intercourse. The view that emphasizes procreation determined the major thrust of Pius XI's teaching concerning sexual activity within marriage. Pius described the act of sexual intercourse itself as primarily destined for the begetting of children and only secondarily aimed to provide "mutual aid, the cultivation of mutual love, and the quieting of concupiscence" (Paragraph 60). Nevertheless, the express toleration of a couple's decision to engage in sexual intercourse even when conception cannot be expected – read by some commentators at the time as tacit approval of the rhythm method – represented an implicit admission of the value of sexual intercourse sought for the sake of expressing love apart from an intention to beget children (Paragraph 60).[8]

A stronger commitment to marital sexuality as an expression of loving, interpersonal unity was manifest in the section of the Vatican II document, *The Church in the Modern World,* which deals with marriage and the family. The assembled bishops declared that marital love is "uniquely expressed and perfected through the marital act." Sexual intercourse of husband and wife "signify and promote that mutual self-giving by which spouses enrich each other" (Paragraph 49). The language of primary and secondary ends of marriage was abandoned in this document. Instead, the document presented a clear portrait of marriage as an institution in which begetting children and promoting interpersonal intimacy between spouses are coequal aims. Church leaders did acknowledge that, in some circumstances, responsible couples might decide to limit the birth of more children. Yet, if abstinence from sexual intercourse is demanded of the married couple, sexual fidelity and the loving nurture of children already born may be imperiled (Paragraph 51). Still, in one of the contradictory impulses in the Council's documents, the bishops also insisted that there can be no fundamental conflict between the desire for genital contact as an expression of love and the ability to conceive children responsibly. Relying on natural law language, the bishops boldly asserted that there can be no fundamental contradiction between the divine laws pertaining to procreation and those pertaining to authentic conjugal love (Paragraph 51).

A continued emphasis on loving unity and procreation as the (inseparable) fundamental aims of marital intercourse is found in *Humanae Vitae,* the papal encyclical that reaffirmed the condemnation of the use of artificial means of birth control. The irony is that substantial *historical change* in the understanding of marital sexuality is the backdrop against which the encyclical lays claim to be in continuity with prior church teaching. Paul VI presents his letter as based squarely upon consistent ecclesiastical analysis of human sexuality understood as "natural" and, hence, unchanging. He asserts: "That teaching [about contraception], *often set forward by the magisterium,* is founded upon the inseparable connection willed by God and unable to be broken by man on his own initiative, between the two meanings of the conjugal act: *the unitive meaning* and the procreative meaning" (Paragraph 12, emphasis added). The value assigned to "the unitive meaning" of marital intercourse constitutes an important new emphasis influenced by historical developments in the modern period.

Currently, the magisterium of the Roman Catholic church seeks to maintain a combination of procreation and interpersonal unity as the two coequal principal aims of marriage. Liberal Catholic theologians, meanwhile, have come to place their primary emphasis on the capacity of sexual intercourse to serve as an expression of profound interpersonal intimacy between partners. The quality of the love expressed in intercourse becomes a crucial element in evaluating the moral legitimacy of sexual activity, at least between marital partners. Lisa Sowle Cahill provides a good summary of this position when she characterizes contemporary Christian sexual ethics as having "permitted procreation gradually to cede primacy of place to the interpersonal aspects of sexuality .... A sexual act may or may not be procreative, but it is always an avenue of personal communication, and a constituent of the most intense and intimate human relationships possible" ([7], p. 141).

This latter normative statement – sexual activity is morally legitimate only when it is an expression of intense interpersonal intimacy – is a culturally based ideal. I suggest that an ideal of intense interpersonal intimacy between heterosexual partners is foreign to the experience of many Christians who have lived in other historical periods. (It may be class bound as well.) I question whether Christians who were socialized in groups in which there were rigid, highly polarized sex roles expected marriage to provide opportunities for profound emotional intimacy. In certain historical and cultural settings, *same-sex* kinship or friendship relationships were an important source of emotional (and, perhaps, sexual) intimacy, particularly for women ([9], pp. 160–96; [41]). To the extent that we recognize explicitly the culture bound nature of the ideal of genital sexual activity as expressive of intensive interpersonal communication, we are free to critique the limitations as well as the benefits of this norm.

The modern recognition of the human person as a being entitled to direct his or her own sexual relationships as well as other social relationships is a genuine moral advance in European and American culture. This personal freedom represents a precious resource for the establishment of sexual relationships characterized by mutual respect and reciprocal self-giving and fulfillment. As a twentieth-century, Euro-American ethicist, I find the belief that sexual intercourse ought to be an expression of reciprocal openness and mutual self-giving very appealing. Still, the cultural vision of a sexual relationship as one of the most powerful expressions of interpersonal love and intimacy is subject to distortion. For example, a facile concept of romantic love can be mistaken for the difficult and fragile process of mutual personal revelation that characterizes genuine interpersonal intimacy. A shallow notion of romantic love lends itself too readily to interpersonal exploitation and irresponsibility. There is a real danger that almost any sexual behavior, no matter how personally or socially destructive, may be condoned because "we really love each other."

In addition, "loving" sexual relationships are not necessarily egalitarian. For example, during the first half of the seventeenth century in England, while

Protestant preachers were stressing the importance of love in a Christian marriage, the patriarchal power of husbands and fathers was increasing and the subordination of women was deepening. Historian Lawrence Stone suggests that these two trends may have been interrelated. "Women were now expected to love and cherish their husbands after marriage and were taught that it was their sacred duty to do so. This love, in those cases where it in fact became internalized and real, made it easier for wives to accept that position of submission to the will of their husbands upon which the preachers were also insisting" ([42], pp. 141–42). Romantic love and patriarchal domination can coexist and can even be mutually reinforcing.

THE SEARCH FOR A USABLE TRADITION

Implicit in my call for collaborative scholarship that would permit us to integrate the historical experiences of women and less ecclesiastically powerful men into the tradition is the hope that such an enriched version of tradition would provide greater resources for a contemporary ethic supportive of sexual wholeness for diverse women. There is enormous work to be done by historians and ethicists before we will know whether that hope can be fulfilled. The search for a "usable history"[9] requires unusual scholarly creativity because the traces of women's historical experiences are often so scanty. Also, it is only certain women – often economically privileged women and, in the Catholic religious tradition, often celibate women – who have had the opportunity to create records reflecting their self-understanding.

Moreover, within recorded historical memory, women have lacked experiences free of influence from their *patriarchal* cultural and social context. For example, although Caroline Bynum wants to emphasize the personal and social meanings that saintly women attached to the practice of rigorous fasting, she also acknowledges that at least some of these women could not transcend entirely the hatred of the female body prevalent in certain aspects of their culture ([5], pp. 216–18). For some of these women, fasting was also a way to chastise their dangerously carnal, female bodies. Thus, I must ask if I can select certain elements of medieval women's piety – those that are consistent with my intuitions about sexual and personal wholeness – and ignore the other parts that seem to threaten an ethic that cherishes the body?

Examining the bodily piety of late medieval women and the resistance against mandatory celibacy has shown me the difficulties of a search for a usable past. Can I be confident that further collaborative scholarship will discover a genuinely "usable" history? Will we find remnants of our foremothers' resistance to male dominance and of women's efforts at self-direction of their lives, particularly their sexual choices? Are there traces of the experiences of female ancestors who valued equality in their sexual relationships? If such memories are uncovered, how do I make ethically useful connections with those memories given the distinctiveness of the experience of women who lived centuries ago in very different social and cultural circumstances?

As human beings, we always carry the past with us in the present. In ethical reflection, as in the rest of life, human beings, individually and corporately, are always reinterpreting and reappropriating elements from the past to move forward in the present. We cannot avoid being selective, for we can never recreate the entire social matrix of some revered period in the past. It seems to me that there is no better alternative than to be honest and self-critical about our reappropriation of various traditions and to be morally accountable to others for the consequences of our use of selected segments of the tradition.

TOWARD A NEW ETHIC OF EROTIC EQUALITY

At the beginning of this essay, I spoke about the silence of women throughout the Catholic tradition. Roman Catholic women today must demand an active role in shaping and sustaining Catholic moral traditions. I suggest that an active and influential role for diverse Catholic women would lead to a radical reorientation of Roman Catholicism's sexual ethics. The task confronting contemporary Catholics is to create a sexual ethic consistent with a deep and abiding regard for the worth and dignity of every human person. The Catholic tradition needs to be reshaped, so that it can contribute to the creation of social and cultural conditions that minimize sexual brokenness and support sexual wholeness in our times. We must formulate a sexual ethic that will be consistent with a major, morally positive cultural change now taking place – consistent with the recognition, at last, that women share fully in human dignity. We need to begin to develop a sexual ethic for a culture, which, in the tantalizing words of philosopher Mariana Valverde, "eroticizes equality" ([44], p. 43).

In this final section, I will discuss three major concerns which need to be addressed in a renewed Roman Catholic sexual ethic. First, Catholic ethicists must work out the implications of a shift away from an emphasis on procreation as the *telos* which justifies sexual activity to an understanding of procreation as one possible good among others which may be realized through (some) sexual relationships. Second, Catholic ethicists need to name violent or coercive sexual activity, not nonprocreative sexual activity, as the fundamental sexual evil. They should examine the implications of unequal distributions of social power for a sexual ethic rooted in respect for the equal dignity of sexual partners. Third, a renewed sexual ethic should consider not just the restraint of sexual evil, but also the enhancement of sexual satisfaction and sexual intimacy as important components of human fulfillment for many persons.

REASSESSING THE HUMAN MEANING OF PROCREATION

Roman Catholic ethics must finally come to grips with a profound and deep seated historical change in the human meaning of procreation. Despite a vigorous rear guard action by the Vatican and certain conservative theologians, many Catholic ethicists and most sexually active Catholics have moved procreation away from the center of sexual morality. Procreation is no longer an unambiguous good, if it ever was. I question whether women who could imagine themselves becoming pregnant would *ever* have developed a sexual ethic that centered on procreation as the primary (unambiguously positive) end of marital intercourse. For women, especially in earlier centuries, multiple pregnancies and births too often represented a serious threat to their survival and that of children whom they had already borne. Many women understand that some pregnancies are extraordinarily burdensome and dangerous and that some infants cannot be offered even the bare minimum conditions for dignified human life.

I suggest that – for a longer historical time than is often recognized – many of the Catholic faithful have realized that they have a moral responsibility to employ effective means to limit births for the good of their family relationships. Catholic couples have been making successful attempts to limit fertility for over two hundred years. By the second half of the eighteenth century in France, for example, the ordinary faithful in large numbers were limiting births and the birth rate declined steeply. Within a century the same trend could be observed in other Catholic nations of Europe ([36], pp. 461–70). Urbanization, advanced industrial economic arrangements, and ecological strains have so transformed social relationships that, barring a sudden, catastrophic depopulation of the globe, responsible and effective control of births is a major moral obligation for heterosexually active, fertile persons.

Women, in particular, are redefining the place of procreation as a moral value in their lives. Women are asserting that an ability to control their own fertility is essential to women's equal participation in other aspects of communal life. This struggle to redefine the meaning of procreation as a good for women may not be a recent historical phenomenon either. Women may have been struggling to come to a new understanding of their procreative responsibilities at prior moments in history in ways ethicists do not yet fully appreciate. For example, John Noonan discusses a Catholic moral treatise addressed to the literate laity of Renaissance Italy. Its author, Cherubino of Siena, castigates women who engage in contraceptive practices to avoid the burdens of pregnancy and motherhood. Noonan comments: "That this motive strikes him as a probable one suggests that already in the late fifteenth century the women of the Italian bourgeoisie were restive in being given only a maternal role" ([36], p. 412). Further scholarly work might reveal to ethicists that "ordinary" Catholic women have long been struggling to assume greater autonomous control over their fertility, precisely in order to act as responsible moral agents. Most women today continue to bear children and many women would describe their maternal relationships as among the most central ones in their lives. Nonetheless, women are asserting ever more strongly that an ability to direct their own reproductive power is essential for their well-being and that of their families.

If the Catholic Church finally put procreation in perspective as an important aspect of human sexuality, but no longer a feature determinative of the moral legitimacy of every act of intercourse, then the church might be able to offer a needed counter cultural witness demanding *social accountability* for the use of procreative power. Fertile, sexually active, heterosexuals do have a serious moral responsibility for new lives which result from their sexual actions. A renewed Catholic sexual ethic would hold such persons accountable to the community for responsible use of procreative power. Such people should be challenged to consider whether they (and their communities) would be able to provide adequate care for a child who might be conceived as a result of their sexual union. If not, they should not have intercourse without employing a reliable means of birth control.

The present Catholic teaching exaggerates the moral significance of procreation for a fertile, married couple; and, perhaps, it underestimates the importance of procreation for those who cannot achieve it through a nuptial embrace. The normative view of sexuality promulgated by the magisterium of the Catholic church extols the interrelated values of love and procreation said to be present in the marital act. Among the problems with that paradigm – as an exclusive ethical reference point – is that it does not provide a helpful basis upon which to develop ethical guidance for infertile heterosexual couples or homosexual persons who wish to become parents. We need to make a fresh examination of ways in which such persons can responsibly be connected to procreation and/or to the nurture of children.

A displacement of procreation from a central position in Catholic ethics will require a reevaluation of the moral weight that we attach to loving companionship and to bodily pleasure in the moral evaluation of sexual behavior. A crucial asset here would be a serious dialogue with gay Catholics, especially lesbian women, whose sexual experiences have been so severely silenced. Some lesbian Catholics have challenging questions to ask about the moral assessment of women's sexuality when it is not defined by men in the service of men and about genuine acceptance of *women's* erotic fulfillment. These are questions for all women, not just those who have genital relations with other women.

NONCOERCION, JUSTICE AND SEXUAL ETHICS

The fundamental evil in a revised code of Roman Catholic sexual ethics should not be nonprocreative sexual activity; rather it should be coercive sexual activity. Attention to the harsh reality of sexual violence puts the attractive ideal of the unitive meaning of human sexuality into a larger perspective. Sexual intercourse can be used as an expression of anger or dominance that destroys any possibility of establishing real interpersonal intimacy. Ethicists need to consider the implications for sexual morality of the social science findings that "sexual assaults by *intimates,* including husbands, are by far the most common type of rape" ([15], p. 7, emphasis added).

Several times in this essay, I have used the metaphor of "sexual wholeness." Our lived experience, as women and men, is an experience of sexual brokenness and betrayal as well as sexual healing or wholeness. A revised Catholic sexual ethic would seek to restrain the evil of coercive or violent sexual activity. This may seem a banal normative statement. Of course, coercive sexual activity is morally wrong. But a firm ethic of noncoercive sexual activity is essential, because, in the United States, all women, all children, and an unknown number of men are at risk of sexual exploitation and sexual violence.[10] Sexual coercion – rape (including marital rape), child sexual abuse, or sexual harassment – represents a morally reprehensible expression of hatred or dominance. It is true that "sexual abuse is only peripherally about sex. More often it [is] ... about humiliation, degradation, anger, and resentment" ([15], p. 18). Still sexual abuse reveals starkly the broken and evil aspects of our sexuality.

One especially powerful reminder of sexuality deformed by human brokenness is marital rape. By marital rape, I mean the use of violence or the threat of violence by a husband (or ex-husband) to compel his wife to engage in sexual activity against her will. Social science studies show that 10 to 14 percent of married or previously married women report that their husbands used force or threat of force to try to make them submit to sexual activity ([15], pp 6–7; [39], p. 57; [18], pp. 542–43).

The Catholic tradition has even contributed to social blindness toward the reality of marital rape. It did so by defining marriage as a religious contract that created a special conjugal right (*jus ad corpus*), i.e., an exclusive, perdurable right to the sexual use of the partner's body. Each spouse was obligated to render "the marital debt," i. e. have sexual intercourse at the partner's request. In fairness, it should be noted that the wife was excused from her marital duty if intercourse posed a serious threat to her well-being or that of the children or if her husband had become an unfit partner. Classic examples of conditions that nullify a husband's right to sexual intercourse with his wife are insanity or drunkenness. Besides, terms such as "the marital debt" have rarely been used by Catholic ethicists since Vatican II.

Still, the Vatican II ideal of sexual intercourse as an expression of loving intimacy is a hollow mockery without explicit insistence on a prior principle of noncoercive sexual activity. A renewed Catholic sexual ethic must speak clearly and frequently against coercive or violent sexual behavior. Sexual behavior which expresses dominance or hatred must be condemned as a violation of sexual equality.

An adequate sexual ethic – one consistent with a principle of equality between the sexes – must include wide-ranging social analysis. In particular, it would explore the extent to which a lack of gender justice in the economic and social realms is a threat to sexual morality. A complex interrelated system of social structures still supports male privilege and deprives women of equal social power, particularly equal economic power. Because of the unequal social, economic, and cultural power held by women and men *as members of their respective gender groups*, it is very difficult for individual heterosexual partners to achieve genuine equality in any aspect of their lives, including the erotic dimension.

Social injustice undermines an ethic rooted in regard for the equal human dignity of all persons in other ways that are seldom explored in discussions of Catholic sexual ethics. For example, sociological patterns and cultural images exert significant influence over our opportunities to form and sustain positive sexual relationships. We do not all have equal opportunities to form personally fulfilling and interpersonally enriching sexual relationships. Homophobic prejudice and institutional heterosexism reverberate in damaging ways in both heterosexual and homosexual relationships. As Mariana Valverde perceptively notes, so-called sexual liberation is a possibility primarily for "people ... who are young, economically secure, and without major responsibilities to children or to anyone else" ([44], p. 188). Opportunities for life-enhancing sexual relationships are too often denied persons who fail to conform to societal norms for physical attractiveness or who have noticeable physical or mental impairments. Racial oppression and ethnic prejudice create conditions that introduce special strains into the relationships between some sexual partners. Unjust patterns for allocating social power distort the distribution of opportunities for personally fulfilling and interpersonally enriching sexual relationships.

AN ETHIC WHICH ENHANCES THE GOOD OF SEXUAL FULFILLMENT

A renewed Catholic ethic should define coercive sexual activity as the fundamental evil and analyze social patterns that undermine the equal dignity of sexual partners. Yet, a renewed sexual ethics should not overemphasize the restraint of sexual evil. Rather it should see an important role for ethical guidance which enhances prospects that human beings will experience the goods of sexual satisfaction and sexual intimacy.

Throughout the "official" Roman Catholic tradition, there is a tendency to view sexual desire as a ferocious form of energy that perpetually threatens to surge out of control with profoundly damaging consequences personally, socially and spiritually. This is too simplistic. Sexual feelings and sexual touch are sometimes more tentative and fragile stirrings of a longing for personal affirmation, solace, intimacy or bodily enjoyment. Sexual desire can be severely and lamentably dampened by low self-esteem, interpersonal rejection, despair created by various forms of social oppression, or even mere physical exhaustion. Ethicist Margaret Farley offers wise advice when she reminds us that moral norms for sexual behavior should not focus exclusively on controlling sexual feelings and activities. "Sexuality is of such importance in human life, and in interpersonal relationships, that it needs to be freed, nurtured, and sustained, as well as disciplined, channeled and controlled" ([14], p. 1587). Other Catholic ethicists need to begin to imagine what it would mean to articulate an ethic that frees, nurtures and sustains sexual energy for the sake of personal wholeness and for the sake of the enrichment of our communal live.

One part of a sexual ethic that nurtures and sustains sexual wholeness would be an enriched notion of the virtue of [sexual] fidelity. The types of commitment between sexual partners that make for human wholeness would need to be reconsidered. The image of faithfulness as a life-long pledge of physical exclusivity in sexual relations is not enough. Faithfulness is a promise to sustain and deepen a relationship in which one is known and valued and knows and values the other. In western culture, women, both gay and straight, have been socialized to integrate our physical desires with our emotional attachments. The yearning to integrate sexual release with loving tenderness is a woman's truth that is worth preserving. (Although women need to be more honest with themselves about whether they sometimes feel sexual desire that is not connected to love.)

Sexual fidelity is a matter of making a promise to oneself and to the other that binds us to the on-going work of loving, rather than simply being carried away by "love." Sometimes the exciting, deeply satisfying quality of sexual joining and the fierceness of erotic passion make it possible for partners to maintain a sound relationship despite hard times. The fidelity that has moral significance is an on-going commitment to be present to one another in all our brokenness and our power. Such commitment is manifest in a steadfast concern for the beloved and for the relationship that allows us to "hold fast when love is not always returned as we had hoped, or when we would have wished," when it "engulfs us far more radically than we had envisaged and stretches us beyond what we had understood as our limits" ([20], p. 189).

If we viewed our sexual energies as a resource to be cherished, we might more easily recognize positive connections between our sexuality and our spirituality. Certain feminists are carefully examining the connection between sexual energy and an even more inclusive "spiritual" force in human experience. (For example: [26]; [34], pp. 180–195; [38], pp. 197–210.) Audre Lorde's essay, "Uses of the Erotic: the Erotic as Power," is an eloquent exposition of these insights. By the erotic, Lorde means a type of energy that fuels genuine sexual passion but also underlies other passionate human feelings and commitments as well. The erotic, according to Lorde, is a sense of internal satisfaction that, once known, challenges us to strive continually to meet our full potential in our personal and corporate lives. The power of the erotic also creates the ability to feel acutely and deeply in the midst of our daily activities. In particular, it unleashes our capacity for joy. Being in touch with the erotic allows us to come together in mutually respectful ways "to share our joy in the satisfying ... [to] make connection with our similarities and our differences" ([30], p. 59). Optimally, the erotic is thread through our work lives, our political action, and our spiritual experiences as well as our sexual couplings.

Lorde does not deny that the power of the erotic can be misused, but she claims that repression of the erotic is not an effective way to prevent possible abuses.

> We have been raised to fear the yes within ourselves, our deepest cravings. But, once recognized, those which do not enhance our future lose their power and can be altered. The fear of our desires keeps them suspect and indiscriminately powerful .... The fear that we cannot grow beyond whatever distortions we may find within ourselves keeps us docile and loyal and obedient, externally defined, and leads us to accept many facets of our oppression as women ([30], pp. 57–58).

When we begin to shape our lives in terms of the erotic, i.e., our own inner knowledge of what gives us joy, we become unwilling to endure powerlessness and oppression passively. As Lorde says: "Recognizing the power of the erotic in our lives can give us the energy to pursue genuine change within our world, rather than merely settling for a shift of characters in the same weary drama" ([30], p. 59). Lorde connects our sexual longings to a deeper, more inclusive yearning for all that is good in human life. As a religious ethicist, I would suggest that this hunger is finally a hunger for contact with divine power.

This way of connecting the sexual in all its bodily intensity with the spiritual has enormous appeal for me. There is a truth here – a truth, I sometimes, but only sometimes, know through my experience of erotic desire. Still, the theme of the positive power of the erotic is recurring often enough in religious feminist discourse that it is time to begin to explore its possible negative facets as well as its obviously attractive aspects. Let me illustrate a few of the questions feminists might begin to explore. Are there nonfeminist cultural sources for this view of the erotic, as well as wellsprings in our own most satisfying personal experiences? Is this celebration of the erotic a feminist rendering – perhaps an unwitting one – of a pervasive, "popular culture" version of Freudian theory? Is it a new way of saying that the sex drive is one of the strongest and most fundamental human urges – one that permeates life, particularly all attempts at cultural creativity? If so, does this idea carry with it any implicit Freudian "baggage" that we would do well to examine and possibly discard? Does the feminist celebration of the erotic have any important similarities to nineteenth-century Romanticism? If so, is it vulnerable to some of the corruptions that plagued the Romantic movement? In suggesting such questions, for which I have no ready answers, I am also suggesting that a feminist ethic of the erotic needs to include an ongoing process of self-criticism.

Such self-criticism is also an important aspect of Catholic women's responsible participation in the process of shaping tradition as members of their church. This struggle of women to become active shapers of sexual morality is a crucial struggle, not just for women's growth in sexual wholeness, *but also for the welfare of the Catholic community*. The entire church – male and female, gay and straight, sexually active and celibate – must engage in a candid reappraisal of the church's teachings about women and human sexuality. If such an honest, far reaching dialogue among a wide range of participants whose insights are truly respected is not given priority soon, the vitality of the church itself will be jeopardized.

Symbolic of the urgency of this task is one recent result of a continuing requirement that ordained ministers be celibate, coupled with an inveterate insistence on an all-male priesthood. There is a serious and rapidly worsening shortage of Roman Catholic priests throughout the world. The shortage of clergy has become so severe, that, in the United States, Catholic bishops are beginning to plan carefully for a time in the not-distant future when many Catholic congregations have no priests available to preside at the Eucharist weekly.

The Sunday Eucharistic liturgy is a central activity of the Roman Catholic community. It is a powerful sign of the unity of believers in the one body of Christ. As a sacrament, it is a sign which effects what it signifies. It creates the bonds of community that it so eloquently symbolizes. Yet, a fear of women and a distrust of sexuality are so deeply ingrained in the personalities of some church leaders, and in certain of the traditions that the bishops are pledged to uphold, that they are willing to jeopardize the celebration of the Eucharist, rather than to allow the body and blood of Christ to be consecrated by any woman or by a man who publicly acknowledges an active sexual life.

If we accept the model of Catholic tradition as a compendium of the sayings of celibate, male theologians, then "the tradition" offers limited resources with which morally serious contemporary women can help men to define an ethic that promotes both sexual wholeness for believers and vitality for the religious community. Now is the time to begin a long-range, collaborative, scholarly project that may allow Catholics to discover a more usable history. Then tradition would include the stories of believing women struggling to deal responsibly with their sexuality.

Catholic women need to contribute to a new sexual ethic, one that is consistent with recognition of the equal worth and dignity of men and women. Such an ethic would recognize that coercive sexual activity, not nonprocreative sexual activity, is the fundamental evil. Ethicists influenced by feminist thought should reexamine the connections between the erotic and a "spiritual" awareness of a divine power that intends wholeness in human lives.

Women need to become full partners in a fundamental reassessment of Roman Catholic sexual ethics. For, unless contemporary Catholicism can come to a better understanding of human sexuality, even its Eucharistic unity may jeopardized. If Catholics – sexually active and celibate, gay and straight, women and men – do not find a way to reorient the tradition where sexuality is concerned, Roman Catholicism's viability as a sensual, sacramental church is threatened.

Monmouth College
West Long Branch, New Jersey, U.S.A.

NOTES

[1] This particular version of the double standard in which women, who are seen as purer and less passionate beings, are held to a higher standard of behavior is a relatively recent cultural

image. At other periods in Western history, women were kept under strict control (at least in the upper classes), not only to protect men from lust, but also to provide external control over women who were viewed as more carnal, less rational, and more prone to sexual promiscuity.

[2] There is a vigorous debate underway in contemporary Roman Catholic ethics about the proper understanding of natural law as an ethical concept. In this paragraph and the next one, I refer to a classicist conception of the natural law. A classicist view purports to offer access to moral norms which are certain and unchanging. In the area of human sexuality, norms are frequently apprehended by reflecting on the physical structures of the reproductive organs and by contemplating intercourse as a biological activity. Careful consideration is given to the generative faculties that human beings share with other animals. This view of the natural law continues to be influential, particularly in documents concerning human sexuality issued by the Vatican. For more information about differing conceptions of natural law, see ([21], chapters 15 and 16), [11], and ([37], chapters 13 and 14).

[3] Certain male mystics have also used erotic imagery to describe their experiences as persons beloved by Christ.

[4] Those who opposed mandatory clerical celibacy are "losers" from the point of view of the official Roman Catholic magisterium which still requires celibacy of its ordained ministers. Protestant church historians who come from Christian traditions that have rejected clerical celibacy have a different perspective on this history. See, for example, [1].

[5] This is not to say that loving affection between husband and wife, especially as it might develop during the course of the marriage, was not recognized as a human good. Some writers both ancient and medieval praised such marital affection. For example, Thomas Aquinas spoke of the great friendship between husband and wife. Such sweet bonds develop among human beings, in part, because of the pleasurable unity of copulation (*Summa Contra Gentiles,* III, 123). Yet, this is the same theologian who also said that woman was created to be man's partner in procreation, since in any other human endeavor a man will find another man to be a superior companion (*Summa Theologiae,* I, 98). Loving friendship between husband and wife was recognized by some premodern writers as a wonderful side benefit of matrimony, but the theologies of the church did not recognize it as a central purpose of either marriage or marital intercourse.

[6] If this historical analysis of the growth of medical and psychological professionals' power to influence sexual norms is correct, it raises ethical questions for such professionals. How can this cultural power be used in a socially responsible way?

[7] Throughout this essay I will use the term magisterium to refer to the teaching authority of the popes and bishops who, by virtue of their episcopal office, have a special charge to preserve the integrity of the Roman Catholic church's beliefs concerning faith and morals. Some Catholic theologians have also used the term magisterium to speak of a different type of teaching authority in the church exercised by theologians by virtue of their scholarly competence. In this essay magisterium is not used to speak of the contribution of theologians. For further discussions of the two "magisteria," see [13]. For a general treatment of the magisterium, see [43].

[8] A similarly complex understanding of nonprocreative, marital sexual activity is reflected in the church's practice of permitting marriage (or continuing sexual activity) when one or both spouses are infertile.

[9] Letty Russell used this term, "usable history," in a pioneering work of feminist theology [40] .

[10] There is a high incidence of rape. Given the prevalence of rape, all women can be described as at risk of being raped. There is a high incidence of child sexual abuse, with some indications that male children are victimized more often than had originally been understood. Therefore, all children can be described as being at risk of sexual abuse. Sexual abuse of adult males does occur. However it has not often been investigated. It is hard to estimate the risk of sexual abuse faced by nonincarcerated adult men.

BIBLIOGRAPHY

1. Bailey, D.S.: 1959, *Sexual Relation in Christian Thought,* Harper and Brothers, New York.
2. Barstow, A.: 1982, *Married Priests and the Reforming Papacy: The Eleventh-Century Debates,* Edwin Mellen Press, New York.
3. Bayer, E.: 1985, *Rape Within Marriage: A Moral Analysis Delayed,* University Press of America, Lanham, Maryland.
4. Brown, P.: 1988, *The Body and Society: Men, Women, and Sexual Renunciation in Early Christianity,* Columbia University Press, New York.
5. Bynum, C.: 1987, *Holy Feast and Holy Fast: The Religious Significance of Food to Medieval Women,* University of California Press, Berkeley.
6. Cahill, L.: 1986, 'Sexual Ethics', in J. Childress and J. Macquarrie (eds.), *The Westminster Dictionary of Christian Ethics,* Westminster Press, Philadelphia, pp. 579–583.
7. Cahill, L.: 1985, *Between the Sexes: Foundations for a Christian Ethics of Sexuality,* Fortress Press, Philadelphia.
8. Church, F.: 1975, 'Sex and Salvation in Tertullian', *Harvard Theological Review* 68, 83–101.
9. Cott, N.: 1977, *The Bonds of Womanhood: Woman's Sphere in New England, 1780–1835,* Yale University Press, New Haven.
10. Countryman, L.W.: 1988, *Dirt, Greed, and Sex: Sexual Ethics in the New Testament and Their Implications for Today,* Fortress Press, Philadelphia.
11. Curran, C.: 1970, 'Natural Law and Contemporary Moral Theology', in *Contemporary Problems in Moral Theology,* Fides Publishers, Notre Dame, IN, pp. 97–158.
12. D'Emilio, J. and Freeman, E.: 1988, *Intimate Matters: A History of Sexuality in America,* Harper and Row, Philadelphia.
13. Dulles, A.: 1980, 'The Two Magisteria: An Interim Reflection', *Catholic Theological Society of America, Proceedings* 35, 155–69.
14. Farley, M.: 1978, 'Sexual Ethics', in W. Reich (ed.), *Encyclopedia of Bioethics,* Free Press, New York, 4, pp. 1575–1589.
15. Finkelhor, D. and Yllo, K.: 1985, *License to Rape: Sexual Abuse of Wives,* Holt, Rinehart, and Winston, New York.
16. Fiorenza, E.: 1984, *In Memory of Her: A Feminist Theological Reconstruction of Christian Origins,* Crossroad, New York.
17. Fitzgerald, A. (trans.): 1926, *The Letters of Synesius of Cyrene,* Oxford University Press, London.
18. Frieze, I.: 1983, 'Investigating the Causes and Consequences of Marital Rape', *Signs: A Journal of Women in Culture and Society* 8, 532–53.
19. Furstenberg, F. et al.: 1987, 'Parental Participation and Children's Well-Being after Marital Dissolution', *American Sociological Review* 52, 695–701.
20. Gudorf, C.: 1985, 'Parenting, Mutual Love, and Sacrifice', in B. Andolsen, C. Gudorf, and M. Pellauer (eds.), *Women's Consciousness, Women's Conscience: A Reader in Feminist Ethics,* Winston Press, Minneapolis, pp. 175–91.
21. Gula, R.: 1989, *Reason Informed by Faith: Foundations of Catholic Morality,* Paulist Press, Mahwah, NJ.
22. Harrison, B.: 1983, *Our Right to Choose: Toward a New Ethic of Abortion,* Beacon Press, Boston.
23. Harrison, B.: 1985, *Making the Connections: Essays in Feminist Social Ethics,* ed. C. Robb, Beacon Press, Boston.
24. Hart, C. (trans.): 1980, *Hadewijch: The Complete Works,* Paulist Press, New York.
25. Herman, D.: 1989, 'The Rape Culture', in J. Freeman (ed.), *Women: A Feminist Perspective,* Mayfield Publishing Company, Mountain View, CA, pp. 20–44.

26. Heyward, C.: 1984, *Our Passion for Justice: Images of Power, Sexuality, and Liberation,* Pilgrim Press, New York.
27. Heyward, C. et al.: 1986, 'Lesbianism and Feminist Theology', *Journal of Feminist Studies in Religion* 2, 95–106.
28. Lebacqz, K.: 1990, 'Love Your Enemy: Sex, Power, and Christian Ethics', in D. M. Yeager (ed.), *Annual of the Society of Christian Ethics,* Georgetown University Press, Washington, DC, pp. 3–24.
29. Lewin, T.: 1990, 'Father's Vanishing Act Called Common Drama', *New York Times,* 4 June, A18.
30. Lorde, A.: 1984, 'Uses of the Erotic: The Erotic as Power', in *Sister Outsider,* Crossing Press, Trumansburg, New York, pp. 53–59.
31. Miles, M.: 1988, *Practicing Christianity: Critical Perspectives for an Embodied Spirituality,* Crossroad, New York.
32. Milhaven, J. G.: 1989, 'A Medieval Lesson in Bodily Knowing: Women's Experience and Men's Thought', *Journal of the American Academy of Religion* 57, 341–72.
33. Mitterauer, M. and Sieder, R.: 1982, *The European Family: Patriarchy to Partnership from the Middle Ages to the Present,* University of Chicago Press, Chicago.
34. Mud Flower Collective: 1985, *God's Fierce Whimsy: Christian Feminism and Theological Education,* Pilgrim Press, New York.
35. Nelson, J.: 1978, *Embodiment: An Approach to Sexuality and Christian Theology,* Augsburg Publishing House, Minneapolis.
36. Noonan, J.: 1965, *Contraception: A History of Its Treatment by the Catholic Theologians and Canonists,* New American Library, New York.
37. O'Connell, T.: 1978, *Principles for a Catholic Morality,* Seabury Press, New York.
38. Plaskow, J.: 1990, *Standing Again at Sinai: Judaism from a Feminist Perspective,* Harper and Row, San Francisco.
39. Russell, D.: 1982, *Rape in Marriage,* Macmillan Publishing Company, New York.
40. Russell, L.: 1974, *Human Liberation in a Feminist Perspective – a Theology,* Westminster Press, Philadelphia.
41. Smith-Rosenberg, C.: 1975, 'The Female World of Love and Ritual: Relations between Women in Nineteenth-Century America', *Signs: A Journal of Women in Culture and Society* 1, 1–29.
42. Stone, L.: 1977, *The Family, Sex, and Marriage in England 1500–1800,* abridged ed., Harper Colophon Books, New York.
43. Sullivan, F.: 1983, *Magisterium: Teaching Authority in the Church,* Paulist Press, New York.
44. Valverde, M.: 1985, *Sex, Power and Pleasure,* The Women's Press, Toronto.

GEORGE EDWARDS

CONTEMPT OR COMMUNION? BIBLICAL THEOLOGY FACES HOMOSEXUAL INCLUSION

The question of ecclesial and synagogal inclusion[1] of uncloseted lesbians and gay men places upon Hebrew-Christian biblical understanding a task of revolutionary ramifications the implications of which have only begun to be felt. We face the challenge of a historic shift of biblical paradigms, and no guarantee exists that teachers of scripture will successfully meet the challenge, carrying their communities forward into a new place of promise or leaving them in the sad bondage of rancorous controversy.

In the ensuing discussion of biblical ideas it is assumed not only that the text itself is culture-bound, but also the perceptual power of the one who expounds the text.[2] Arriving at a biblically guided morality freed from the culture-boundedness of the text is a difficult enterprise indeed. To cite a simple example, angels are not in contemporary thinking a commonplace phenomenon. Furthermore, is it not impossible, in the final analysis, to overcome the culture-boundedness of the interpreter of the text?

Yes, it is impossible in any total sense of the task, yet the interpreter must struggle to subject one's comprehension to evaluation and reform. The worst of circumstances is to reject either of these formidable interpretative responsibilities. The text in itself must be critically understood, and the reader's or the community's way of grasping it must also be critically tested, since no one can boast of presuppositions that need no reform.

In the effort to determine how our biblical traditions contribute to or detract from sexual wholeness and healthy sexual functioning, with special reference to the subject of homosexuality, various ways of dealing with the two facets of comprehending the text are evidenced in the following sections of this essay. My reading of Genesis 19, for example, holds that tradition – a historical stream of understanding claiming its basis in the text – is clouded over with a cultural homophobia that obscures the larger redemptive motif of hospitality[3] made accessible through critical interpretation. Secondly, the datedness of the text itself is demonstrated in its lack of a language (e.g., the absence of the word homosexuality itself in Hebrew and Greek) now normative in scientific discourse, as well as the means of verifying empirically the various assumptions related to same-sex orientation or practice.

I

Genesis 19:1–11 is a good beginning point for bringing into focus our need for a new understanding of a familiar biblical paradigm.

Ronald M. Green (ed.), Religion and Sexual Health, 79–97.

The angels of Jahweh, after royal treatment (Genesis 18) by Abraham and Sarah at Mamre, proceed (Genesis 19) to Sodom. Lot's gracious reception echoes that at Mamre. But at nightfall, the wicked city gathers at the door as the stage is ominously set for the raping of the divine visitors. Only a miraculous intervention stands between them and a brutal humiliation. The impingement of this violent tale upon the theological, moral, legal and medical enterprises through the word "sodomy" and its cognates poses in the sharpest way the dilemma of contempt or communion our contemporary attitudes and conduct face. In a series of itemized reflections on this passage that follow, it will be clear by deduction what the traditional path of understanding has been, as well as that for which I am appealing. For each item a minimum of elaboration is given.

- Genesis 19:1–11 is a tale of wholesale masculine aggression devoid of all qualities of love expressed in mature homosexual relationships.
- Judges 19:18–30, an unmistakable literary parallel to the passage in Genesis 19, demonstrates in heterosexual terms the gang violence that makes up the common moral denominator in the two narratives and shows that heterogenital sexual crime is the same as homogenital sexual crime.
- The literary unity of Genesis 18 and 19 shows their central moral motif is built on the hospitality/inhospitality axis. This is equally conspicuous in the Judges 19 parallel.
- Hermann Gunkel, the pioneer of critical study of Genesis in this century, has shown the solid rootage of Genesis 19 in ancient folklore where an unrecognized deity (or deities) visits but is inhospitably received by nearly all the inhabitants of some ancient place. This inhospitality brings on a catastrophic destruction from which the hospitable ones are miraculously delivered.[4]
- The gross inhospitality that marks Sodom in Genesis 19 comes out distinctly in the prophetic use of the Sodom tradition as shown from Ezekiel 16:49, "Behold, this was the guilt of your sister Sodom: she and her daughters had pride, surfeit of food, and prosperous ease, but did not aid the poor and needy."
- The "outcry" against Sodom cited in Genesis 18:20, 21, and 19:13 becomes that which signals the city's coming doom. According to von Rad ([37], p. 211) the technical term for outcry (*ts*^{e}*aqah*) has to do with the screams raised up by those who have been the victims of violent injustice. It is not an outcry about having the wrong kind of sex.
- The hospitality/inhospitality theme is the only descriptive content given to Sodom in the tradition of Jesus's sayings. The texts are Matthew 10:15, 11:24; Luke 10:12, and the Majority text of Mark 6:11. These presumably emerge from the Q source – this source refers in general to the material common to Matthew and Luke but lacking in Mark and have nothing to do with homosexuality.
- The traditional view of Genesis 19 gives no consideration to three important aspects of contemporary sexological study: the difference between same-sex

orientation and same-sex practice; the non-pathological quality of homosexuality; and the increasing tendency among embryologists, developmental psychologists, and sexologists to regard sexual orientation as a congenital or prenatal rather than an acquired characteristic.[5]
- The total masculinity of Genesis 19:4 highlights the moral questions raised by contemporary society in response to such pictures of pervasive patriarchy and justifies the complaint that not even a single woman of the city rates mention as a lesbian. That astonishing text reads, "The men of the city, the men of Sodom,[6] both young and old, all the people to the last man, surround the house."
- If all the men of Sodom are gay, why does Lot have sons-in-law as indicated in 19:12 and 14? Are he and his daughters ignorant of the notorious sexuality of their city?
- The phrase, "both young and old," in 19:14 is an all-inclusive age span. The Hebrew for "young" (*na*ᶜ*ar*) is found in Exodus 2:6 as also in Judges 13:5 to refer to a newborn baby. And how old is old? If one excludes these tiny tots from the impending assault on the angels of Jahweh, why are they included in the fire and brimstone holocaust soon to follow, especially if, as some have maintained, Genesis 18:22–33 is an example of divine compassion in dealing with Sodom?
- From the standpoint of biblical criticism in our time,[7] it must be recognized that the fiery destruction of Sodom and its inundation at the southern end of the Dead Sea has no confirmatory archaeological evidence behind it. Even if such evidence should eventually appear, that is not confirmation of the theology traditionally connected with Genesis 19, namely, the angry God who, except for Lot, his hapless wife, and two daughters, burns up every last man, woman, and child from the youngest to the oldest, on account of the wrong kind of sex.

II

In this section it will be shown how the traditional understanding of the destruction of Sodom opposed in part I has been used in the interpretation of Genesis 1–3 and read into those primeval accounts of creation in order to portray same-sex orientation and homogenital conduct as a defection from a created, natural, paradisic heterosexuality, and a byproduct of original sin.

By drawing attention now to the 1986 "Letter to the Bishops of the Catholic Church on the Pastoral Care of Homosexual Persons," my purpose is not to single out with customary Protestant disapproval the errancy of Catholic moral and biblical teaching. Indeed, the central features of that teaching will be found to coincide with those of my own Presbyterian (U.S.A.) denomination.

Section 5 of the Letter to the Bishops[8] correctly emphasizes the "varied patterns of thought and expression" which mark biblical literature. This correct premise of the "remarkable diversity" in scripture then yields to the questionable assertion that "there is nevertheless a clear consistency with the Scriptures themselves on the moral use of homosexual behavior." Solidarity with this "constant biblical testimony," section 5 continues, affirms that "the Scriptures are not properly understood when they are interpreted in a way which contradicts the Church's living tradition. To be correct, the interpretation of Scripture must be in substantial accord with that tradition."[9] The circularity of the Vatican rationale is clear. Tradition has arrived at a singularly negative judgment on what the Bible says about homosexual behavior, and other interpretations must comply with that. Tradition exercises the right to tell us what scripture says and then to rule out alternative views.

"Homosexual behavior," furthermore, is a net which catches too many fish. All conduct of homosexual people can be called homosexual behavior, if a reductionist tendency, like the popular view of Genesis 19 encourages, permits a whole class of people to be seen from a single genital perspective. Though the Vatican letter uses the expression "homosexual orientation" three times (sections 3, 11, and 16), it disinclines from any positive attributions to it such as are often found in scientific studies of sex and in religious publications.

The letter prefers terms like condition, inclination, and tendency. Further, "the inclination itself must be seen as an objective disorder" (section 3). Robert Nugent ([31], pp. 52–53) cannot then be corrected when saying that the Vatican gives only theoretical recognition to the difference between orientation and behavior while equating homosexuality and homogenital behavior and thus equating also orientation and homogenital behavior. We can doubt whether this avoids the reductionist tendency mentioned above. While "violent malice in speech or action" is deplored in section 10 it is also true that even carefully laundered language can be a vehicle of contempt. Whoever utters such language cannot disclaim responsibility for the violence it produces.

Failure to give informed attention to twentieth century developments in the understanding of sexual orientation invites the misconception that sexual preference[10] is simply a matter of the individual's decision.

The same problem seriously affects the interpretation of Romans 1:26–27, "Their women exchanged natural relations for unnatural, and the men likewise gave up natural relations with women and were consumed with passion for one another, men committing shameless acts with men ..." This can only refer to persons who violate a natural heterosexual orientation in order to engage in homogenital practices. The text assumes an entirely different coloration when one concedes the actuality of a homosexual orientation. In this case, same-sex acts are not a violation of the constitutional predisposition of the person. In section 6, however, the Vatican letter, having disregarded the very aspects of human sexuality that cast new light on ancient texts, reiterates the time-worn condemnation said to emerge from Genesis 19, as though the millennia that have passed since the writing of these two passages had brought about on the essential points not a fractional modification of linguistic, religious, or moral understanding.

As to language, it is important that "homosexual" is a relatively modern word and did not exist in biblical Hebrew, Aramaic, or Greek. It came over into English only in 1892 by translation of Krafft-Ebing's German *Homosexualität*. This explains the absence of homosexual or homosexuality in the King James Version of 1611. So the King James translators resorted to the word sodomite in translating the Hebrew *qadesh* at Deuteronomy 23:17; 1 Kings 14:24, 15:12, 22:46, and 2 Kings 23:7, encouraging thereby the contemptuous equation: homosexual equals sodomite.

Homosexual people are not sodomites. As a class, they are not characterized by gang rape any more than gang rape is a characteristic of heterosexual people because of what we read about the thugs of Gibeah in Judges 19.

When the Vatican letter (section 4) attacks the notion that scripture "somehow tacitly approves of it," i.e., homosexuality, the fulcrum of the issue is: what kind of homosexuality? It is a delusion to think that homosexual persons approve of or practice what is described in Genesis 19. Lesbians and gay men do have some acquaintance with rape. This has to do, almost without exception, with instances in which they have been victims of rape perpetrated by heterosexuals.

When section 4 remonstrates further against the falsity of a new exegesis that claims all moral injunctions of the Bible are "so culture-bound that they are no longer applicable to contemporary life," it is describing a straw man. It is certainly undeniable that the Bible in many morally important respects is culture-bound. The institution of slavery as such is not condemned. Slaveholders are required to treat their slaves "fairly and justly" (Colossians 4:1/Ephesians 6:9), but that is a far cry from emancipation.[11] Androcentricity, despite a few qualifications here and there, is reinforced from Genesis to Revelation.

In the various aspects of human sexuality, Catholic teaching is particularly vulnerable to antiquarian cultural vestiges. Anticontraceptive precepts based on the "be fruitful and multiply" of Genesis 1:28 are grossly at odds with the practice of millions of Catholic people moved by the economic and social realities of modern parenting as well as the moral factors of global demographics. Celibacy doctrines derived from "eunuchs for the sake of the Kingdom of Heaven" (Matthew 9:12) and the sexual abstinence of Jesus as dealt with in church tradition now contribute significantly to the dearth of candidates causing the closure of increasing numbers of Catholic seminaries while sacramental ministries are correspondingly neglected or abandoned. The alienation of Catholic women ready and educationally prepared for priestly ordination is a ticking bomb the destructive power of which is hidden only to those bound to patriarchal culture with the help of biblical prooftexts and church tradition carefully formulated by men only.

Genesis 19, as could be expected, also plays a vital role in the logic of the Vatican letter. Part of section 6 reads as follows:

In Genesis 3, we find that this truth about persons being an image of God has been obscured by original sin. There inevitably follows a loss of awareness of the covenantal character of the union these persons had with God and with each other. The human body retains its "spousal significance" but this is now clouded by sin. Thus, in Genesis 19:1–11, the deterioration due to sin continues in the story of the men of Sodom. There can be no doubt of the moral judgment made there against homosexual relations.

Here Sodom, fall, and original sin are brought together to form a solid phalanx against modifications that might be proposed on the basis of factors itemized in part I of this discussion. The similarity of this to the United Presbyterian judgment against the ordination of "practicing homosexuals" in 1978 can now be described.

A two year study of this ordination question was authorized by the United Presbyterian Assembly of 1976. The group appointed for this study was made up of 19 persons. After two years of intensive inquiry, reflection, biblical study, and prayer, 14 took the affirmative position: lesbians and gay men having met the usual requirements of ordination could be ordained as Presbyterian ministers without a vow of celibacy.

The majority report was written by Byron Shafer [40] a specialist in biblical studies at Fordham University in New York. The majority report contains serious expositions of the relevant biblical passages. It also draws upon information current to 1978 about scientific contributions to sexology, for example, the decision of the American Psychiatric Association in 1973 and that of the American Psychological Association in 1975 to remove homosexuality from the index of psychic pathology.

Favoring ordination within the committee of 19 was a majority of almost 3 to 1, but when the San Diego Assembly of 1978 voted on the majority report (composed by Shafer), it was rejected by a proportion of 9 to 1.

Subsequent ruling of the Permanent Judicial Commission of the denomination (known since 1983 as the Presbyterian Church (U.S.A.)) has given the negative decision of the '78 Assembly the effective status of an amendment to the church's constitution, without observing the process for such an amendment, namely, a confirmatory vote among a majority of the presbyteries.[12]

The 1978 language of the United Presbyterians is only slightly less brazen than that of the '86 Letter to Bishops cited above. Aware perhaps that Genesis 1–3 says absolutely nothing about homosexuality, the Presbyterian Policy Statement ([36], p. 261c) adopted by the Assembly majority bridges the gap between Genesis 19 and Genesis 3 by asserting, "Genesis offers polemic against deviations from the wise separation of humankind into man and woman." The context makes it clear that the word "deviations" can only refer to homosexuality.

The language of the fall continues to guide the Presbyterian doctrine: "one explanation of the process in which persons develop homosexual preferences and behavior is that men and women *fall away* [emphasis added] from their intended being because of distorted or insufficient belief in who they are" (p. 262a). The fall and gayness are linked again in words like this: "Even where the homosexual orientation has not been consciously sought or chosen, it is neither a gift from God nor a condition like race; it is a result of our living in a fallen world" (p. 262b). Redemption and heterosexuality are clearly joined in the context of the statement that Christ calls us "out of the alienation, brokenness, and isolation of our fallen state into the freedom of new life in Christ" (p. 262c). Still again Presbyterians and Catholics echo each other in this Presbyterian pronouncement:

> As we examine the whole framework of teaching bearing upon sexuality from Genesis onward, we find that homosexuality is a contradiction of God's wise and beautiful pattern for human sexual relationships revealed in Scripture ... (p. 262d)

Richard Lovelace, Presbyterian Professor of Church History at Gordon-Conwell Seminary, was a primary spokesperson for the minority of the 1978 study group. His book on *Homosexuality and the Church* ([24], p. 103) asserts that "The starting point for understanding both human sexuality in general and homosexuality should be the account of the creation of man and woman in Genesis 1 and 2." But in his discussion of "The Biblical Evidence," in chapter 4 of his book, the sequence of attention moves from Leviticus 18:22 and 20:13 to Romans 1:26–27 and other New Testament texts, then back to Genesis 19 and Judges 19, before his arrival at the creation narratives.

It is clear from this that Lovelace's procedure is to lay down first what the Vatican letter calls "a clear consistency within the Scriptures themselves on the moral issue of homosexual behavior." Lovelace's words for this are what he calls "a central, pervasive, and consistent body of doctrine running through the whole structure of biblical ethics" ([24], p. 103).

But if the creation texts are the "starting point," it should not be necessary to invert the whole order of the biblical stories, so the reason for this must now be clear. The basis for the old ecclesial contempt must first be reiterated and then read back into Genesis 1–3. This excursion is necessary in order that, with more than a little help from Augustine's idea of original sin and its refurbishment by John Milton, heterosexuality can be pictured as paradise and homosexuality as paradise lost.

The incredibility of this excursion should also be altogether obvious. If the fall of Genesis 3 is in some way related to the fall from heterosexuality into homosexuality, why do the various editors of these ancient traditions make us wait 15 chapters before we find out what the fall was all about? Hence the great leap from the disobedience of Adam and Eve to the sodomy of the Sodomites is not a credible leap forward but a return to the silence which provoked the great leap in the first place. This extended hiatus is, on the other hand, made understandable on the basis of the critical reading of Genesis 19 proposed in section I of this chapter.[13]

III

Earlier I defended the view that when Paul speaks of giving up or exchanging "natural relations for unnatural" in Romans 1:26–27, it could only refer to persons who violate a natural heterosexual orientation in order to engage in homogenital practices and cannot refer to persons whose homosexual orientation is to them natural. This view of this important New Testament text is in general accord with that expressed by John Boswell ([9], p. 109), namely, "Paul did not discuss gay *persons* but only homosexual *acts* committed by heterosexual persons."[14]

When the United Presbyterian Policy Statement ([36], p. 262b) refers to Romans 1:26f and asserts that "By 'unnatural' the Scripture does not mean contrary to custom, nor contrary to the preference of a particular person, but rather contrary to that order of universal human sexual nature that God intended in Genesis 1 and 2," it stigmatizes homosexual persons by placing them outside the intention of God in creation. It is noteworthy, furthermore, that the Policy Statement in treating Romans 1:26–27 follows the very same pattern as that previously described in its linking of Genesis 1–3 and Genesis 19. That is, Romans 1:26f is presumed to be a condemnation of homosexual behavior, and this condemnation goes back to the creation stories which imply the condemnation already established in the correlative text. Thus, Romans 1:26–27 explains Genesis 1–3 and vice versa.

Richard Hays ([18], p. 191) is so enamored with this idea that he imagines Paul portraying "homosexual behavior as a 'sacrament' (so to speak) of the anti-religion of human beings who refuse to honor God as creator: it is an outward and visible sign of an inward and spiritual reality, figuring forth through 'the dishonoring of their bodies' the spiritual condition of those who have 'exchanged the truth about God for a lie' (1:24–25)." As if this did not already express enough contempt, Hays continues: "Thus, Paul's choice of homosexuality as an illustration of human depravity is not merely random: it serves his rhetorical purpose by providing a vivid *image* of humanity's primal rejection of the sovereignty of God the creator."

In the first part of his statement, Hays is restrained enough to designate homosexual *behavior* (as over against the state of being homosexual), but in the latter part of the statement, this attention to the text is exchanged for the word "homosexuality," abandoning the clinical differentiation between acts and orientation basic to the interpretation of Boswell.

So the question must now be posed, does Romans 1:26–27 rest in actuality on the heterosexual features of the creation in Genesis 1 and 2 and the fall from that (into homosexuality) represented in Hays's exegesis and the intended meaning of Paul in Romans 1:18–32? In my view the answer to this is no.

1. To be sure, creation (*ktisis*) and Creator (*ho ktisas*) are found in Romans 1:20 and 25.[15] The anti-idol polemic of Romans 1 is also strongly marked by the infinite qualitative difference between the Maker and that which is made. Genesis 1:26–31, however, emphasizes the glory of humankind made in the

image of Elohim – Psalm 8:5 similarly praises human beings as "a little less than God" – and accents the goodness of all created things. The divergent emphasis in Romans 1:18–32 arises from the forceful attack on Gentile depravity specifically manifested in the idol making of those who refuse to acknowledge the glorious deity of the Creator and express befitting thanks. The motifs of this pervasive theme have repeated parallels in the Wisdom of Solomon (Wisdom, for short) 13–15, plus a few Old Testament texts that do not include Genesis 1–3. Wisdom is an apocryphal book (i.e., not contained in the Hebrew canon of the Old Testament nor Jerome's Vulgate) written in Greek and arising from Alexandrian Judaism in the first or second century before Christ. Chapters 13–15 are an untiring, often satirical attack on the futility of idol making and idol worship, "the beginning of fornication" (14:12) and a Pandora's box of all other Gentile wrongdoing. Commentaries substantiate with minute detail the parallels between Romans 1:19–32 and Wisdom 13–15.[16]

It is also helpful to point out at this juncture that the 26th edition of the Nestle-Aland Greek Testament, the standard international resource for study of the Greek New Testament, begins its marginal notation on Romans 1:19–32 by listing for comprehensive comparison Wisdom 13–15. In its subsequent references (dozens of them) not one mention of Genesis 1, 2, or 3 is to be found. These references are not endowed with *ex cathedra* finality, but neither should their complete omission of any Genesis parallels be trivialized.

The usual division of Romans 1:18–32 from 2:1ff holds that Gentiles are the idolaters under judgment in the first case,[17] while the chosen people suffer the same judgment in 2:1ff. This perspective is understandable in view of the attitude toward idol worship in Wisdom. The folly of Canaanite idolaters, who will get what is coming to them despite God's longsuffering, is expounded in Wisdom 12:3–11, but the major attention in Wisdom seems to focus on the idolatry of Egypt. This frame of reference doubtlessly shapes the "they"/"them" pronouns of Romans 1:18–32, and underlies the sudden emergence of "you"/"your" which appear in Romans 2 and are explicitly identified in 2:17 as referring to the Jew.[18]

But the Gentile/Jew sequence (Paul's normal usage is Jew/Gentile cf. Romans 2:9, 10; 3:9, 29) is compromised by another consideration germane to Paul's rhetorical subtlety. That is, even the section on Gentile idolatry turns out to be addressed to Israel, for at 1:23 Paul takes up the language of Psalm 106:20, the episode of the golden calf described in Exodus 32. This spoils, so to speak, the "them and us" frame of thought suggested by conventional reading of Romans 1:18-32. Markus Barth[19] and Klaus Berger ([3], p. 26) have discerned the anomaly that results when the last half of Romans 1 is taken to refer to Gentiles. As Berger remarks, "Of course 1:23 refers unmistakably to the golden calf, and the Test. Naphtali 3–4 is to be compared to the entire text, where it expressly says of the Jews who can recognize the creator (!), that they will do the evil of the heathen. In the wickedness of the heathen, according to Rom. 1:18ff, Jews and heathen are the same."

To recapitulate, Genesis 1–3 does not underlie Paul's thought in the last 15 verses of Romans 1. Paul's way of approaching the sin subject has more subtlety than that. This is intimated by the reference to Hebrew idolatry in a section ostensibly engaged with Gentile idolatry, resulting in an anomaly. The solution to this lies in the fact that Paul's letter is controlled not by a general premise having to do with a universal doctrine of sin but by his resolve to establish his Gentile apostolate in the face of Jerusalem's resistance to it. This resistance and exclusion threaten the validity of Paul's entire mission.

The moral correlative to this within the concerns of human sexuality is clear. The contempt with which the heterosexual majority views the homosexual minority rests on a similar self righteous "them and us" outlook. Just as faith righteousness or the righteousness of God (Romans 1:17) renders invalid all legal righteousness based on some human achievement or qualification, Paul shapes his message in Romans 1 to subvert the boasting of his Jerusalem adversaries and make possible dignity and communion for all.

In the following concluding section of my argument, confirmations of this attitude toward exclusivity will be adduced, showing that ecclesiastical and civil homophobia cannot correctly claim biblical sanction.

IV

To use Romans 1:26f as a basis for the penalization of persons of homosexual orientation or those who engage in homosexual acts – when adequate attention to the quality of the relationships between the persons so involved is not paid – violates the spirit and intent on which the whole of Romans is premised.

In his work on *Paul Among Jews and Gentiles,* Stendahl ([42], p. 2) has opened a window of understanding on Paul's unyielding defense of faith justification as over against a righteousness based on the law. In regard to this preeminent issue of Romans, Stendahl holds that "the doctrine of Justification by faith was hammered out by Paul for the very specific and limited purpose of *defending the rights of Gentile converts to be full and genuine heirs to the promises of God to Israel*" (emphasis added). This clearly expresses the sociological reality which lies underneath the theological affirmation for which Paul is best known. Galatians confirms this. When Peter under pressure from James in Jerusalem withdraws himself (Galatians 2:11–14) from table fellowship or communion with Gentile converts, Peter, as H.D. Betz avows ([4], p. 30), "wants to make them into converts of Judaism. This contradicts the principles of the doctrine of justification by faith, which had been the basis of the faith thus far (see 2:15–16)."

The implications of this for sexual inclusivity are obvious. When Paul pronounces the baptismal solidarity of Galatians 3:27–28, he is affirming a communion in which human accreditations and distinctions (neither Jew nor Greek; neither slave nor free; neither male nor female) have lost their weight. This nondiscriminate, egalitarian inclusiveness is not only germane to but indispensable for a contextual understanding of Romans 1:26–27.

If we ask the question, why then did Paul not say in Galatians or Romans "neither gay nor straight?," the answer is not hard to find. The gay/straight language and even the word homosexual itself did not exist in either Hebrew or Greek in Paul's day, as we previously noted. The latter term, translated from German, does not appear in English until 1892.

Robin Scroggs ([39], pp. 116–117) may be, furthermore, correct in saying that Paul when referring to same-sex practices has only pederasty in mind, because that form of same-sex behavior was familiar to both Greek and Roman civilization and fell into disrepute not alone in Judaism and Christianity but also among certain Greco-Roman moralists because of the abuse or potential abuse of minors. In note 14, I indicated the contextual grounds on which Romans 1:26f may have to do with sacral prostitution, male and female, even though the very meager evidence does not permit us to know exactly what these temple functionaries did.

Finally, the social and moral tasks of biblical religion can never be confined to literal descriptions of them contained in biblical books. What does scripture teach about atomic waste disposal? About in vitro fertilization? About government deregulation? About the veil (1 Corinthians 11:5ff) or women speaking in church (1 Corinthians 14:34–36)? Time makes the list longer. History erodes moral judgments and requires of us new ones if the word of God is to remain vital, present, and revelatory. Fundamentalism seeks to maintain the authority and worth of biblical texts by refusing to recognize the relativizations of history. This refusal, despite its commendable intent, can only result in defeating the very purposes which gave rise to it. This applies, moreover, not to Christianity and Judaism alone, but to other religions and the cultures they produce or affect.

Romans 15:25 discloses that Paul is "going to Jerusalem with aid for the saints." Macedonia and Greece (15:26) have taken up these collections, and Paul is the agent for their delivery. 1 Corinthians 16 and 2 Corinthians 8 and 9 also give instructions concerning the offerings. The usual inference is that Paul writes Romans before his departure from Corinth for Jerusalem or during the time of the journey itself.

Bornkamm ([7], pp. 88–96; [8], pp. 17–60) fixed the writing during Paul's last winter in Corinth, 55–56 AD, and provided meaningful insight into the character of the Roman epistle.

Paul is filled with serious forebodings about the impending encounters in Jerusalem. He asks (15:30) for supportive prayers from the Roman community in burdened language (cf. *sunagonizesthai*), lacking even the assurance that the offerings from the Gentile congregations will be accepted (15:31b).

In Acts 20:17–38, Luke's account of Paul's speech to the Ephesian elders reports (vs. 23) that imprisonment and affliction are waiting for him in the holy city. This is reinforced in Acts 21:10–14 when a prophet, Agabus, at Caesarea foretells, in acted parable, of Paul's coming imprisonment. His friends beg him to stay away from Jerusalem.

Although Paul wants to get to Spain by way of Rome (15:24, 29), fulfilling his longstanding desire to visit Rome (1:10–15), the Roman letter has the quality of a general defense of Paul's mission to the Gentiles and the faith righteousness which we encounter in the other letters of Paul.[20] This gives Romans some features of a last will and testament, without being a compendium of Paul's whole religious thought. The letter expresses, nevertheless, with striking fullness the faith righteousness or righteousness of God (cf. Romans 1:17) and the Gentile inclusion giving rise to it, as argued by Stendahl. Furthermore, it is the impending Jerusalem clash[21] far more than the needs or troubles of the Roman community, which focuses Paul's feelings. So much is this so, that Jacob Jervell ([19], pp. 61–64) can write of Romans as "The Letter to Jerusalem."

Justification by faith is the foundation of human dignity in contest with every form of exclusion and contempt expressed in structures of privilege and power. Righteousness from God[22] brings into question every humanly derived righteousness based on nationality, class, race, gender, or sexual orientation. From the religious and moral standpoint, it makes all arrogance ridiculous (Romans 3:27–31) and calls into communion all those cast down or cast out by the politics of contempt. This pervasive, overarching theme of Paul's testament makes it certain that Romans 1:26–27 is wrongly understood if it eclipses or contradicts Romans' foundational inclusivity. When these two verses are used to encourage the belief that homosexual orientation or practice (when it involves mature, loving adults in a relationship committed to constancy) is sinful, a fault of one's own or one's parents, it is a form of sexual chauvinism based on a purely human credential, a contradiction of the righteousness of God and the communion that righteousness enables and sustains.

Ernst Käsemann ([20], pp. 53–54) notes that the transition to Romans 2 from the ending of Romans 1 "can be understood only as a polemic against the Jewish tradition which comes out most clearly and with much the same vocabulary in Wisdom 15:1ff," which reads:

> But thou, our God, art kind and true, patient and ruling all things in mercy. For even if we sin we are thine, knowing thy power; but we will not sin, because we know that we are accounted thine. For to know thee is complete righteousness, and to know thy power is the root of immortality. For the evil intent of human art has not misled us.

The sharp turn in Romans 2 to the confidence of the boastful elect now makes it clear that the upshot of the preceding ideas in 1:18–32 is not, as often claimed, an expression of Paul's own moral repugnance at Gentile depravity, but, by reciting the stereotypic attitude of diaspora Judaism looking out upon the idolatrous corruption of the surrounding Gentile world, Paul now withdraws from it every ground of substantiation. You, the pious one, are condemned by your own fatal self confidence. You preach but don't practice.

This is also the reason that in the section (1:18–32) usually assigned to God's wrath on Gentile idolatry, Israel's own idolatry is a telling part of Paul's exhibit of evidence.

Such rhetorical reversal is as old as Nathan's indictment of David in 2

Samuel 11:1–12:7. After arousing David's righteous indignation by a story about an unjust man, Nathan announces: "you are the man" (2 Samuel 12:7). The consequence of this rhetorical tour de force is clear. Romans 1:18–32 is not contextually a passage expressing Paul's moral admonition (technically speaking, parenesis) but a glimpse of Paul's attitude toward the impending Jerusalem confrontation.

This should not be understood to mean that Paul wants to acquit the Gentile world of all wrongdoing. At the same moment, he will not endorse the chauvinism which would require of Gentile converts, to whom his missionary efforts have been so sacrificially given, a form of piety that would subject these converts to the "works of the law." As Brevard Childs ([10], p. 256) correctly observes, "the righteousness of God" which originates in the Old Testament, does not "mean God's revenging or distributive justice, but God's saving righteousness which [God] establishes in relation to [God's] people."[23] Paul did indeed, as Childs holds, "greatly sharpen the term and developed it polemically in the debate with Judaism ..." Romans 1 and 2 become in the manner previously described in this discussion a primary example of this sharpening and polemicization.

The ramifications of this for the theology of Paul and the sexual ideas to be derived from Romans 1:26–27 are now disclosed. The attitude of most mainline religious communions, despite the encouraging exceptions discussed in note 1 require that lesbians and gay men become in manners and attitudes carbon copies of the heterosexual world before they can be religiously acceptable, first class members of established congregations. Some denominations would even consider extending to them, however, the right of ordination, as long as they maintained celibacy and did not openly speak of their sexual orientation.

The more homophobic groups define ministry among gay people as convincing them that Jesus or God can turn them into heterosexuals. The Seventh Day Adventists followed this procedure with Colin Cook who gained considerable publicity by appearances on the Donahue show, vigorously affirming his conversion to a straight life and quoting well memorized biblical texts substantiating the miraculous reality of this change as well as the godforsakenness of those following the gay way. Cook's "Quest Learning Center" financed by the Adventist denomination proceeded to apply Cook's and the denomination's ideas of biblical sexuality in the counseling of homosexual people who entered their program. Six years later Quest Learning Center was closed after reports from clients that Mr. Cook was using counseling sessions – I do not know the frequency of these incidents – for more than verbal expression of his homosexual preference. The denouement of this story is found in an interview with Cook by J. Robert Spangler [41] in the Adventist journal, *Ministry*, which discloses that the denomination sees no fault in its biblical understanding but only in Cook's weakness of faith. Cook concurs in this.

While "recovery" is sometimes effected in cases of bisexuality falling in the medium range of Kinsey's scale of orientational gradations,[24] such changeovers, when counseled by those embracing homophobic attitudes derived from biblical texts, misunderstand the righteousness of God and the socially transforming inclusiveness it creates. This inclusive communion is open to lesbians and gay men in search of the place of intersection between human love and the love of God, apart from the generally futile expectation of a change of orientation.

Louisville Presbyterian Theological Seminary
Louisville, Kentucky, U.S.A.

NOTES

[1] Inclusion is not meant in the narrow sense of closeted membership in a religious community, which has always been more or less accessible to homosexual people. Before its refusal, for example, to allow ordination of "practicing homosexuals" in 1978, the United Presbyterian Church in the USA had already elected to its highest office, the moderator of the General Assembly, a closeted gay man who served admirably in this capacity. The Unitarian-Universalist Church has never officially denied to persons gay or lesbian access to membership or ordination. There are instances of the ordination of uncloseted candidates in the Episcopal Church and the United Church of Christ. The Universal Fellowship of Metropolitan Community Churches with its many lesbian and gay pastors was founded in 1968 by the Rev. Troy Perry. After coming out, he had been divested of ministerial office as a Pentecostal. The United Church of Canada lifted the bar to homosexual ordinands in 1989. Reform Judaism in June 1990 voted to admit acknowledged non-celibate homosexuals to its rabbinate. Affirming the right of ordination has not meant that the practice of specific congregations to vote for or against any candidate as its religious leader has been overruled.

[2] E.V. McKnight [28] gives extensive demonstration of this concern. Pp. 146–151 comment on Paul, Augustine, Luther, Stendahl, and Käsemann in a way pertinent to discussions brought up later in this chapter.

[3] I use hospitality, of course, in the simple sense of the gracious treatment of guests, but the lineage of the term in biblical literature has two additional qualities, depending on the context. 1. These guests may be unrecognized deities or their agents (see Genesis 18:1–2 and 19:1; the great judgment discourse in Matthew 25:31–46, where the messianic Human One is not recognized (vss. 37f, 44) even by the hospitable; Luke 24:28–35; Hebrews 13:2; etc.) 2. In the prophetic tradition, the treatment of the stranger is integral to and indistinguishable from the general laws of social justice, because "you were strangers in the land of Egypt." So the neighbor love of Leviticus 19:18 is the same as that enjoined in the treatment of strangers, Leviticus 19:33–36. This feature is strangely present in Luke's account of the Good Samaritan (Luke 10:29–37) where neighbor love not only includes the alien, but the alien, *mirabile dictu,* becomes the agent of it. Matthew 25:31–46 also represents this compounded subtlety ("I was a stranger").

The depth of hospitality as moral metaphor has been convincingly developed by Thomas Ogletree [33], borrowing from Paul Ricoeur's exposition of a "de-centering of perspective" which seeks to overcome the tendency of much ethical reflection to make the self, its potentiality and actualization, the center of all moral judgments. Here again from a modern standpoint informed by the biblical legacy the alter and the alien serve the same, relatively indistinguishable function in the moral shaping of the ego. Ogletree recognizes (p. 6) that hospitality in the simplex sense (hostess-host graciousness to guest) is limited in its ability to relieve structural social inequity.

[4] The prime example of this is in Ovid ([34] viii, 610–715).

[5] Only a sample of the increasing support for the last of these three assertions can be

described. F.J. Kallmann [21], [22] reported 100% concordance for homosexuality in 37 sets of identical twins even though the co-twins developed their sexuality independently of each other and often far apart. R.H. Bixler [6] pointed out that all the twins of Kallmann's studies denied incest with the co-twin, reinforcing the congenital thesis. Other investigations have not replicated the 100% concordance reported by Kallmann but regularly find "quite high" concordances (Whitam and Diamond [46]) which give strong support to heredity as a major determinant of sexual orientation.

Twin studies were given a larger familial setting by Pillard, Poumadre, and Caretta [35] who found homosexual siblings occur in greater frequency in the families of their probands "than would be expected given population frequencies" (p. 468). Thus family studies confirm a role for heredity.

Heredity is not the only determinant of sexual orientation. Atypical hormonal events can alter sexual orientations. Numerous studies manipulating the prenatal hormonal events as documented in laboratory studies of animals are summarized by Bixler [5], Tourney [44], and especially Ellis and Ames [14]; the latter conclude that "learning, by and large, only appears to alter how, when, and where the orientation is to be expressed" (p. 251). Young, Goy, and Phoenix [50] gave reasoned arguments not only for hormonal factors in the shaping of sexual behavior in experiments with infrahuman vertebrates but also demonstrated the presence of such factors during prenatal development. Dörner [13] and his associates conducted a large number of widely published laboratory experiments on rats and hamsters showing the impact of hormones on brain differentiations and their consequences in sexual orientation and behavior. Dörner's extensive contributions, related to both fetal and postnatal interventions, are summarized in Tourney's ([44], pp. 54–55) study of hormonal elements.

Bell, Weinberg, and Hammersmith [2] also discuss the nurture or nature issue. They cite the classical sexologists, Havelock Ellis and R.v. Krafft-Ebing, in support of a "constitutional or hereditary origin," (p. 212), name (notes to p. 213) five more scientists, previously opposed, who now support this view, and appeal to a prophetic passage from Freud looking to the time when biological factors will dominate the field, while producing extensive evidence earlier in the book against etiological claims based on configurations of the child-mother-father relationship.

Whitam and Mathy [47] hold that "homosexuals are a permanent manifestation of human sexuality found at about the same rate in all societies" (p. xii). Societies do not create homosexuals; they only react to them.

R.H. Bixler [5] to whom I am indebted for significant help in the composition of this note, contends that the denial of heredity as also the denial of environmental affects is only a matter of ideological rhetoric. Neither nature nor nurture should be claimed as the exclusive basis of sexual phenomena.

[6] The textcritical possibility of removing this redundant clause as a gloss would produce only insubstantial improvement.

[7] Cf. T.L. Thompson ([43], p. 174).

[8] Citations from the various numbered sections of the letter are from Gramick and Furey ([16], pp. 1–10).

[9] James Gustafson ([17], p. 4) recognized this feature of Catholic church authority, but with insufficient criticism. Perhaps his object to reconcile Catholic-Protestant disparities transferred power to the magisterial institution of the church, a commendable transfer only if individual decisions of conscience are less responsible than an institutionalized magisterium. My own tendency would be to prefer one Hans Küng, or one Charles Curran, or one Jeannine Gramick to ten Sacred Congregations for the Doctrine of the Faith. John McKenzie ([27], p. 10) assessed with spice the (then) 69 year old Pontifical Biblical Commission when it was submerged in the Sacred Congregation.

[10] Bell, Weinberg, and Hammersmith ([2], p. 222), commenting on their book entitled *Sexual Preference,* warn that they "do not mean to imply that a given sexual orientation is the result of a conscious decision or is as changeable as the many moment-by-moment

decisions we make in our lives. Neither homosexuals nor heterosexuals are what they are by design. Homosexuals, in particular, cannot be dismissed as persons who simply refuse to conform. There is no reason to think it would be easier for homosexual men or women to reverse their sexual orientation than it would bs for heterosexual readers to become predominantly or exclusively homosexuals."

[11] John Knox [23] made a persuasive case that Paul is asking for the freedom of Onesimus to labor with Paul, but if that is correct, it is a salutary exception rather than what we usually find.

[12] If such a vote had been taken, it would probably have sustained the amendment to the constitution of the Presbyterian Church (U.S.A.). A June, 1989 survey of 3,900 Presbyterian members, elders, pastors, and specialized clergy conducted by the Presbyterian Panel (the denomination's official polling agency) reported according to Jerry Van Marter ([45], p. 5-N) 74% of the members and 80% of the elders held that homosexual sex is "always wrong." Only the specialized clergy showed less than 50% (i.e., about 40%) in this category, with pastors at about 55%. Panelists voted more tolerantly if the relationship were viewed as a "loving, caring one." The education of and by the pastors and specialized clergy could, potentially, shift the denomination's attitude. Theological curricula have in the last generation increased emphasis on sex education.

[13] The United Presbyterian Policy Statement of 1978 [36] and that of the Sacred Congregation would command more biblical interest if pursued along source critical lines, since indeed Genesis 2:4b-3 and Genesis 19 have a common Jahwistic denominator. Those who voted for this statement seem as little interested in this approach, however, as they are in the plain divergencies of the two creation stories.

If biblical criticism is ill suited to the bolstering of traditional precepts, administrative control, and church politics in general, that is not a proof of its moral or theological virtue in every instance, however important it may be as leverage against ecclesial power. George Mendenhall [29] refused to affirm Genesis 3 as Jahwistic (i.e., 10th-9th century material), but counted it a parabolic satire against the professional "wise" (cf. Genesis 3:6 *sakal, hiph^cil*), royal counselors, patrons of power and influence, after the collapse of Solomonic wisdom and power in the exilic catastrophe foreseen by the trustworthy preexilic prophets. Mendenhall cites in support of the postexilic date F.W. Winnett [48]. While I share Mendenhall's pessimism about politics and the corruptive power it wields over men and women alike – he is not in favor of male-dominant readings of Genesis 1–3 – he concedes too little to the criticism of biblical (and contemporary) patriarchy.

[14] Boswell does not suggest the probability that he modernizes Paul, that is, ascribes to Paul the appropriate but modern distinction between being homosexual and acting as such. I am not unwilling to concede that my understanding of natural/unnatural in Romans 1:26–27 arises from the validity of reader-oriented criticism. (See note 2 above and the three introductory paragraphs immediately antecedent to section I of this chapter.) "Nature" according to Paul (1 Corinthians 11:14) teaches that long hair (how long?) for a man is degrading (i.e., morally wrong), while the opposite is true for a woman. The passage stumbles because it gets tangled up early on (vs. 3), with a metaphorical use of "head," used patriarchally (vs. 3b, 7b, 8, 9), and is made inconsistent with the insertion of verses 11 and 12. See Conzelmann ([11], pp. 181–191, especially p. 190, and notes). The passage illustrates why Paul's use of "nature" cannot be employed in a morally normative sense but must be subordinated to the freedom authorized by Paul's appeal for faith-based righteousness.

Nor can it be so easily dismissed (cf. Boswell [9], p. 108) that temple prostitution is what Paul has in mind. It is clear that the passage Romans 1:18–32 is suffused with the subject of pagan idolatry. "Exchanged" (*metallassein*) in vs. 26 links back to the "exchanged" in both 26 (*allassein*) and 23 (*metallassein*) and reflects a Judaistic tradition – the context of Leviticus 18:22 and 20:13 indicates the same pattern – wherein idolatry provoking God's "giving them up" (*paradinonai*) in vss. 24, 26, and 28 is the root cause of a general sexual mayhem and a long list of other vices (vss. 29–31). In sum, sacral prostitution and idolatry are closer to hand than a noncontextual scrutiny of Romans 1:26–27 might disclose. The vocabulary of "exchange" is also found in Wisdom 14:26 in a phrase (*geneseōs enallagē*) rendered by David Winston ([49], p. 269), "interchange of sex roles." Others translate "sexual perversion," and commentators who relate this to homosexual persons or acts have not been lacking. Winston (p. 280) cites for comparison Philo, *De Cherubim* 92, and the Testament of Naphtali 3:4. Wisdom 14:12–31 presents another context unmistakably focused on idolatry. Cf. Wisdom 14:12 and 27.

[15] Claromontanus (valuable 6th century Greek and Latin manuscript of the Pauline epistles) also reads *ktisis* for *chrēsis* at vs. 21.

[16] The parallels are printed in adjacent columns in Sanday and Headlam ([38], pp. 51–52). While these parallel columns do not appear in Cranfield [12], he does provide numerous allusions to Wisdom in the verses under inquiry. Nygren ([32], pp. 114–115) substantiates the more than accidental concordance between Romans and Wisdom in this section of Romans and carries his remarks over into Wisdom 15:1 and 2 (at the opening of Romans 2) with much insight. Eduard Grafe ([15], pp. 271–274) argues well the position I advocate, with some updating.

[17] See, for example, the note to 2:1–11 in the *New Oxford Annotated Bible with the Apochrypha* ([30], p. 1362), "*2:1–11: Jews are under judgment,* as well as pagans (1:18–32)." Ernst Käsemann ([20], p. 36) captions Romans 1:18–32, "The Revelation of God's Wrath on the Gentiles."

[18] This "they"/"you" rhetoric was underscored by William Manson ([25], p. 174).

[19] Markus Barth ([1], p. 291) says, "Paul does not speak of Gentiles only in 1:18ff, but of 'the Jews first and the Gentiles' (cf. 1:16; 2:9f)."

[20] Bornkamm ([8], pp. 25–27) itemizes the various themes of Romans and cites the parallels in the previous epistles.

[21] See Acts 21:21 and 28, 28:17.

[22] The genitive in this expression at Romans 1:17 is in fact ablatival, as shown by *hē ek theou dikaiosunē* in Philippians 3:9.

[23] I have used brackets to avoid gender-specific language for God.

[24] Zilbergeld and Evans [51] have pointed out, using Masters and Johnson's book [26] on *Homosexuality in Perspective* and subsequent evaluation of the Masters and Johnson data, that 82% of the patients treated by the St. Louis physicians fall in the middle three categories of the Kinsey orientational scale. This makes it difficult to say the least to think these clients were lesbian or gay rather than bisexual. They also approve (p. 40) of Wardell Pomeroy's complaint that they used the Kinsey Rating Scale only in the behavior sense and not in the arousal sense also, as Kinsey intended.

BIBLIOGRAPHY

1. Barth, M.: 1955, 'Speaking of Sin (Some Interpretive Notes on Romans 1:18–3:20)', *Scottish Journal of Theology* 8, 288–296.
2. Bell, A.P., Weinberg, M.S. and Hammersmith, S.K.: 1981, *Sexual Preference: Its Development in Men and Women,* Indiana University Press, Bloomington.
3. Berger, M.: 1977, *Exegese des Neuen Testaments: Neue Wege vom Text zur Auslegung,* Quelle & Meyer, Heidelberg.
4. Betz, H.D.: 1979, *A Commentary on Paul's Letter to the Churches in Galatia,* Hermeneia, Fortress, Philadelphia.
5. Bixler, R.H.: 1980, 'Nature Versus Nurture: The Timeless Anachronism', *Merrill-Palmer Quarterly of Behavior and Development* 26 (2), 153–159.
6. Bixler, R. H.: 1983, 'Homosexual Twin Incest Avoidance', *Journal of Sex Research* 19, 296–302.
7. Bornkamm, G.: 1971, *Paul,* D.M.G. Stalker (trans.), Harper & Row, New York.

8. Bornkamm, G.: 1977, 'The Letter to the Romans as Paul's Last Will and Testament', in K.P. Donfried (ed.), *The Romans Debate,* Augsburg, Minneapolis, pp. 17–60.
9. Boswell, J.: 1980, *Christianity, Social Tolerance, and Homosexuality,* University of Chicago Press, Chicago.
10. Childs, B.S.: 1984, *The New Testament as Canon: An Introduction,* Fortress, Philadelphia.
11. Conzelmann, H.: 1975, *1 Corinthians,* J.W. Leitch (trans.), Hermeneia, Fortress, Philadelphia.
12. Cranfield, C.E.B.: 1975, *The Epistle of Paul to the Romans,* Vol. 1, International Critical Commentary, T & T Clark, Edinburgh.
13. Dörner, G.: 1976, *Hormones and Brain Differentiation,* Elsevier, New York.
14. Ellis, L. and Ames, A.: 1978, 'Neurological Functioning and Sexual Orientation: A Theory of Homosexuality-Heterosexuality', *Psychological Bulletin* 101 (2), 232–258.
15. Grafe, E.: 1982, 'Das Verhältnis der paulinischen Schriften zur Sapientia Salamonis', in A. Harnack u.a., *Theologische Abhandlungen C.V. Weizsächer gewidmet,* J.C.B. Mohr (P. Siebeck), pp. 251–286.
16. Gramick, J. and Furey, P.: 1988, *The Vatican and Homosexuality,* Crossroad, Chicago.
17. Gustafson, J.M.: 1978, *Protestant and Catholic Ethics,* University of Chicago, Chicago.
18. Hays, R.B.: 1986, 'Relations Natural and Unnatural: A Response to John Boswell's Exegesis of Romans 1', *The Journal of Religious Ethics* 14, 286–215.
19. Jervell, J.: 1977, 'The Letter to Jerusalem', in K.P. Donfried (ed.) *The Romans Debate,* Augsburg, Minneapolis, pp. 61–74.
20. Käsemann, E.: 1980, *Commentary on Romans,* G. Bromiley (trans. & ed.), Wm.B. Eerdmans, Grand Rapids.
21. Kallmann, F.J.: 1952, 'Twin and Sibship Study of Overt Male Homosexuality', *American Journal of Human Genetics* 4, 136–146.
22. Kallman, F.J.: 1952, 'Comparative Twin Study on the Genetic Aspects of Male Homosexuality', *Journal of Nervous and Mental Disease* 115, 283–298.
23. Knox, J.: 1959, *Philemon Among the Letters of Paul,* University of Chicago, Chicago.
24. Lovelace, R.: 1978, *Homosexuality and the Church,* Revell, Old Tappan, New Jersey.
25. Manson, Wm.: 1951, *The Epistle to the Hebrews,* Hodder & Stoughton, London.
26. Masters, W. and Johnson, V.: 1979, *Homosexuality in Perspective,* Little, Brown, New York.
27. Mckenzie, J.L.: 1971, 'No Runs, No Hits, No Errors', *National Catholic Reporter* July 30, 10.
28. McKnight, E.V.: 1988, *Postmodern Uses of the Bible: The Emergence of Reader-oriented Criticism,* Abingdon, Nashville.
29. Mendenhall, G.: 1974, 'The Shady Side of Wisdom: The Date and Purpose of Genesis 3', in H.N. Beam et al. (eds.), *Old Testament Studies in Honor of Jacob M. Myers,* Temple University Press, Philadelphia, pp. 319–334.
30. *The New Oxford Annotated Bible with the Apocrypha,* 1977: Oxford, New York.
31. Nugent, R.: 1988, 'Sexual Orientation in Vatican Thinking', in J. Gramick and P. Furey (eds.), *The Vatican and Homosexuality,* Crossroad, New York, pp. 48–58.
32. Nygren, A.: 1949, *Commentary on Romans,* C. Rasmussen (trans.), Muhlenberg, Philadelphia.
33. Ogletree, T.W.: 1985, *Hospitality to the Stranger,* Fortress, Philadelphia.
34. Ovid: 1971, *Metamorphosis,* Vol. 1, F.J. Miller (trans.), Loeb Classical Library, Harvard University Press, Cambridge, MA.
35. Pillard, R.C., Poumadre, J. and Caretta, R.A.: 1981, 'Is Homosexuality Familial? A Review, Some Data, and a Suggestion', in *Archives of Sexual Behavior* 10 (2), 465–475.
36. Presbyterian Church (U.S.A): 1978, 'Policy Statement and Recommendations' in *Minutes of the General Assembly of the United Presbyterian Church in the U.S.A.,* Part 1, Journal, Office of the General Assembly of the Presbyterian Church (U.S.A),

Louisville, Kentucky, pp. 261–266.
37. von Rad, G.: 1972, *Genesis,* Westminster, Philadelphia.
38. Sanday, W. and Headlam, A.: 1896, *The Epistle to the Romans,* International Critical Commentary, Scribner's, New York.
39. Scroggs, R.: 1983, *The New Testament and Homosexuality,* Fortress, Philadelphia.
40. Shafer, B.E.: 1978, 'The Church and Homosexuality', in *Minutes of the General Assembly of the United Presbyterian Church in the U.S.A.,* Part 1, Journal, Office of the General Assembly of the Presbyterian Church (U.S.A.), Louisville, Kentucky, pp. 213–260.
41. Spangler, J.R. and Cook, C.: 1987, 'Homosexual Recovery – Six Years Later', *Ministry* 69 (9), 4–9.
42. Stendahl, K.: 1976, *Paul among Jews and Gentiles and Other Essays,* Fortress, Philadelphia.
43. Thompson, T.L.: 1974, *The Historicity of the Patriarchal Narratives,* Beiheft zu Zeitschrift für die Alttestamentliche Wissenschaft, 133, Walter de Gruyter, New York.
44. Tourney, G.: 1980, 'Hormones and Homosexuality', in J. Marmor (ed.), *Homosexual Behavior,* Basic Books, New York, pp. 41–58.
45. Van Marter, J.: 1990, 'Three-Quarters Believe That Homosexual Sex is Wrong', *The News of the Presbyterian Church (U.S.A.),* Office of the General Assembly of the Presbyterian Church (U.S.A.), Louisville, Kentucky, March issue, p. 5-N.
46. Whitam, F.L. and Diamond, M.: 1986, 'A Preliminary Report on the Sexual Orientation of Homosexual Twins', Paper presented at the Society for the Scientific Study of Sex, Western Region.
47. Whitam, F.L. and Mathy, R.M.: 1986, *Male Homosexuality in Four Societies: Brazil, Guatemala, The Philippines, and the United States,* Praeger, New York.
48. Winnett, F.W.: 1965, 'Re-examining the Foundations', *Journal of Biblical Literature* 84, 1–9.
49. Winston, D.: 1979, *The Wisdom of Solomon,* The Anchor Bible, Doubleday, Garden City, New York.
50. Young, W.C., Goy, R.W. and Phoenix, C.H., 'Hormones and Sexual Behavior', *Science* 143, 212–218.
51. Zilbergeld, B. and Evans, M.: 1980, 'The Inadequacy of Masters and Johnson', *Psychology Today* 14 (3), 29–34.

CHRISTINE E. GUDORF

WESTERN RELIGION AND THE PATRIARCHAL FAMILY

The Judeo-Christian tradition has helped to shape, legitimate, and even sacralize the model of family characteristic of the western world: the patriarchal family. Though the Judeo-Christian tradition neither created nor exclusively controlled the patriarchal family – which both predated it and was manifested in areas outside the influence of both Judaism and Christianity – neither did the Judeo-Christian tradition merely passively accept the patriarchal family, but instead made it central to the religious tradition itself, and even based religious structures on patriarchal assumptions.

Within Christianity in particular, it is necessary to distinguish between the influence of the overall Christian tradition on the family, and the existence within the Christian gospel of a teaching on family contrary to the dominant patriarchal thrust, a teaching which seems to be rooted in the teachings of Jesus himself and to have exercised a great deal of influence in the first century or two of Christianity. In the New Testament, for example, in addition to the patriarchal household codes found in a number of epistles (especially Eph. 5:22–6:9 and Col. 3:18–4:1) which recapitulated prevailing family structure in the Roman empire and accorded with the patriarchal Jewish norms, there are also a number of countervailing passages. In the Gospels, Jesus opposed the generally accepted primacy of familial duty, especially filial duty, in his refusal to interrupt his teaching to see his mother and brothers (Mk 3:31–35), in his refusal to sanction burying one's father before taking up the duties of discipleship (Lk 9:59–62), and in rebutting the woman who blessed his mother for having birthed him ("Blessed rather are those who hear the word of God and keep it!" Lk 11:27–28). While Jesus never directly contravened the dominant/subordinate relationship prescribed for husbands and wives in patriarchy, he did give many examples extraordinary in his time of respect for women, and he demonstrated support for breaking the stereotypically servant role of women in the home (Mary and Martha, LK 10:38–42). Perhaps the strongest evidence for a New Testament tendency to contravene patriarchy comes from Gal. 3:28 and the examples in Paul's epistles of the leadership roles given to women in the early church, some of whom, like Prisca, shared authority in the church with their husbands ([12], Ch. 5). Yet regardless of the existence of this potential for opposing the patriarchal family within the Christian tradition, from the second century onward such New Testament passages were overshadowed by the development of a patriarchal church that stressed those sections of the New Testament supporting patriarchal relations in the family, such as the household codes.

Today, the passages from the Gospels and the epistles which undermine the

Ronald M. Green (ed.), Religion and Sexual Health, 99–117.

legitimacy of the patriarchal model are being lifted up in some quarters of Christianity as the basis for constructing alternative models of Christian family and community [12], [13], [5]. A similar movement is taking place within Judaism, though the primary texts of Judaism are somewhat less explicitly supportive of alternatives to patriarchy, due largely to their greater antiquity and their original social location [7], [38].

In this discussion I will restrict myself to suggesting some of the effects of three specific aspects of the traditional Judeo-Christian model of family: the headship/breadwinner role of husbands/fathers, wives' subordination to husbands and restriction to motherhood and the domestic hearth, and the imperative that children obey parents. These have been standard aspects in Judeo-Christian tradition, culled from the Genesis story of creation (men as breadwinners, women as childbearers subject to men) and from the commandments (children's obedience to parents), and embodied and elaborated in countless stories in both Hebrew and Christian scriptures. Both Jewish and Christian scripture scholars and religious writers have further discussed, elaborated, and confirmed these three aspects of religious teaching on family throughout history up to and including our century [9]. The effects of these three aspects of the traditional family on both contemporary individuals in the family and on society itself are enormous, and frequently destructive. I will examine the effects of these three aspects on the three groups of participants in the modern family: men, women and children.

EFFECTS ON MEN

The division of roles and power in the traditional family has both privileged and disadvantaged men in varying ways. Since the family has been understood as the primary and basic social unit, the power of men over women and children in the family also served to give men a power, a freedom, in wider society that women, and certainly children, did not have. For example, the chief reason that Thomas Aquinas gives for the impossibility of women becoming priests is that women were created subject to men in families, and for a much narrower purpose (reproduction), so that ordination could not be effective for women *(Summa Theologiae,* III Suppl. Q 39). Similarly Jewish women were not obligated by any of the time bound positive laws – such as on prayer and study of the law – since these would interfere with women's obligations in the home. Because women were ignorant of the law, they were excluded not only as judges, but as witnesses ([19], p. 191). Thus the power of men over women, and their comparative lack of restriction to assigned roles in the family, also conferred on men power outside the family itself. One result of this greater social power and freedom of men has been greater opportunity for personal development and for the formation of identity. Psychologists still find today that men of college age have usually reached a stage in identity formation that most women do not reach before middle age ([45], p. 175).

With social power for men came the power to name and define the world; the material world, the nature of men, women and children, as well as ultimacy itself came to be male-defined. One must be careful not to exaggerate this matter, since rigid hierarchies among men have, throughout history, severely restricted the number of men who exercised social power; for most men, rule over one's own home ("a man's home is his castle") was minimal compensation for exclusion from the male elites who ruled the wider society.

Within the family men's power over women and children tended at many periods in history to be absolute: he determined the place of residence, controlled all family resources regardless of their source, and had the right to use physical force – even to the point of death – to punish women or children or to compel obedience. In some periods and places he could sell them as property, whether in slavery, apprenticeship, or indenture. Within the marital relationship, the greater power of men allowed the sexual double standard, under which women, but not men, were held to celibacy before marriage and fidelity within marriage. Though both later Jewish and all Christian teaching included insistence on chastity, neither tradition rigorously or consistently attacked the double standard, often treating chastity as an impossible ideal for men due to their sexual nature and their social freedom ([33], pp. 227–228; [19], p. 198; [4], pp. 50–52).

To balance this great power of men, the traditional family imposed on men the responsibility of providing material support for the family. Until the modern era most people subsisted from agriculture, in which women and children participated as well. But the responsibility for family support was the husband's. This responsibility was usually onerous, and was made periodically impossible by droughts and floods, wars, epidemics or other social dislocations.

Besides the burden of responsibility for material support, the assignment of work to men, as means of providing family support, has resulted in some serious distortions in men's lives, especially in the contemporary era. Work ideally serves two other purposes in addition to material support. It is human beings' chief method for participating in and contributing to the larger human community. Work also serves as the major activity through which human persons create themselves by learning and developing their talents, encountering and overcoming challenges, and interacting co-operatively with others. In the modern world work is not only very specialized and therefore varied, but chosen by individuals rather than assigned by class or heredity. The emphasis on work as the means to support a man's family, as that which legitimates his participation and role in the family, has overshadowed for many men these other purposes of human work. Many men today do not feel free to choose work through which they can satisfy personal needs or contribute to their societies in ways meaningful to them. Many are trapped in jobs they actively dislike, some in jobs they feel detract from the broader social welfare; they feel obliged to choose work which provides as much material support as possible. Many workers dislike their work so intensely that they understand it as a sacrifice for their families, who in turn owe them respect and obedience – and greater success than the worker himself achieved ([46], Ch. 2; [40], Ch. 7). This is a problem not only among the working class, where basic levels of support are often difficult to achieve, but is also true for many middle class men. Many middle class college men feel pressured to prepare for careers in accounting, law, medicine, finance and engineering, and to squelch any satisfaction they might feel in courses which might otherwise lead them to careers as poets, nurses, forest rangers, social workers or teachers. This is a common source of depression among college and university students. It is, of course, closely intertwined with more general social pressure to measure one's worth by the size of one's paycheck. Working women, socialized to think of their salaries as secondary to their spouse's (pin money), and who normally take their social status (and economic class) from their husband's job rather than their own, have not felt nearly the degree of pressure to choose work for its remuneration alone.

Today the breadwinner role of man in the west faces an additional pressure from the increased numbers of working women. In many countries the majority of women – even the majority of married women with children – now work. The reasons vary, and include career ambitions among more educated women, larger numbers of female headed households and ideological pressure in communist states, but the overwhelming majority of women work due to the inability of a single wage to support a family in adequate comfort.

The phenomenon of women working in large numbers causes anxiety in men at three levels: 1) it undermines the historic economic dependency of women and children on men on which men's authority in the family was based, 2) it forces men to work with women as co-workers, and sometimes with women as bosses, though they have been socialized to view women as subordinates, and 3) it calls into question the value of men's contribution to the family. This last is important, and often overlooked. Women's role in the family is much more secure than men's. Subordination to men and restriction to domestic work has often been difficult, and has sometimes resulted in abuse of women, but the role of mother and homemaintainer is comparatively impervious to becoming anachronistic. The tasks involved in women's role change, but they have never, despite some ideologically motivated attempts at collectivization, either been eliminated or successfully reassigned away from women. Thus economic depression and other social dislocations, for all their disastrous consequences on general family welfare, are much more devastating for men than for women, in that they steal from men the activity on which their identity, their place and authority in the family, as well as their livelihood, are based.

The greater security of women's role in the family has been augmented by the romanticization of the family in the last few centuries, which was aimed at camouflaging the power inequities in the family from a culture increasingly critical of power inequities. This romanticization described women's role in the family as being the heart of the family, as the center of warmth and as mediator of love and communication, as the primary source of nurture. The child was represented as innocent and carefree, protected and cherished. Men's role, however, did not require any romanticization, as men did not require convincing as to the benefits of the patriarchal family; their role was still presented in terms of headship and breadwinner. At times religious treatment of the burden of breadwinning suggested that this onerous burden was men's admission price to the warm refuge of the family hearth – a warmth generated by women and children.[1]

In the present when many women and children no longer are dependent upon men's breadwinning, what do men have to offer as the admission price of membership in the family? The relation between men and the rest of the family in the traditional model was based on economic need, not emotional ties, though emotional ties often developed. Mutual love as constitutive of relationships, though advocated for the Christian community as a whole, was not described as central to the family, but was assumed to be naturally present in the relation between mother and children. In the household codes, for example, the husband was to love, the wife to obey; the father was to refrain from provoking children, the children to obey; the injunctions are complementary, but not mutual.

A major effect of the traditional family on men, then, which is only highlighted by the loss of economic dependency of families on men, is the depiction, and consequent socialization of men as emotionally isolated persons who relate to others primarily through power and provision.[2] Men's historic roles of ruler and breadwinner in the family have never demanded interpersonal nurturance skills, though many men have developed these independently. In dominant/subordinate relations, it is always the subordinates who need to develop sensitivity, tact, communication and solidarity skills, which, under greatly unequal divisions of power, are actually survival skills.

Another explanation often given for the lesser relationality of men is the different conditions under which male children deal with the task of gender identity, which children face in their first and second years ([39], pp. 20–25; [8], pp. 50, 274). Since virtually all child care during these early years is provided by women, boys cannot use the modelling technique which girls use for achieving gender identity. The absence of fathers or other males, and the fact that the young boy's closest attachment is to a female, makes his task of gender identity much more anxiety-ridden. Boys' technique for reaching gender identity is a combination of separation from the mother or mother-substitute, and reliance on reinforcement and cognitive learning. In this explanation, the restriction of child care to women (and the resulting absence of men in child care) is an important factor forcing boys to renounce their earliest intimate relation (with a woman) in favor of more impersonal cognitive learning if they are to feel themselves men. The implications for men's later capacity for intimacy and for later attitudes toward women are potentially serious.

Today we face massive evidence that at social-psychological levels men are increasingly isolated, anxious, and emotionally repressed, and that these conditions are potentially lethal. Their socialization in and for the traditional family has not equipped them in great numbers for child nurturance or for the emotional self-disclosure necessary for the close friendships and mutual, intimate marriages which become more necessary in modern society as more traditional forms of community and intimacy disintegrate under the influence of mobility and urban anonymity. There is a shift in the masculine paradigm now occurring in the West, largely under socio-economic pressure, which emphasizes the co-operative teamwork required of labor in a mechanized age rather than the stoic physical strength and endurance of traditional masculinity ([39], Ch. 10). This shift, toward masculine interpersonal skills and more mutual forms of relationship, which is now in its early stages and is largely limited to middle class workers, is greeted warmly by many women desirous of more mutuality and intimacy in relationships with men. There is no better way to sell women's magazines than headlines promising: "How to Get Your Man to Know and Share His Feelings"; "How Men Can Learn to Parent Well"; "How Spouses Can Be Partners." Yet men's socialization in and for the traditional family is a primary obstacle to this paradigm shift. The situation of many western men today bears eloquent witness to the cliche that it is lonely at the top. Headship and responsibility for breadwinning have deprived many men of some important human experiences, pressuring them to find identity in power and responsibility.

At a social level, the headship of men in the family has been so accepted that there have been few checks on the illegitimate use of men's power in the family. Today we face horrendous statistics regarding men's abuse of women and children, both inside and outside the family. In the U.S. domestic violence occurs in 10–21 percent of all families, with men as the almost exclusive perpetrators of this violence ([32], p. 11; [42], pp. 21–22). Studies show that between 20–48 percent of women are victims of serious sexual assault by males,[3] with a marital rape rate of 14 percent ([42], pp. 57, 64). Over 30 percent of girls and 10 percent of boys are sexually molested [43] – and the overwhelming majority of the offenders are men. 4.5 percent of all girls in this society are sexually abused by their fathers ([44], p. 10). While there are traditional Jewish and Christian teachings on family which condemn such practices, the religious legitimation of the power of men in families conditions some men to understand wives and children as possessions they may use as they wish, thus undermining the bans. Religious traditions have exhibited peculiar blind spots which also promoted abuse. Christianity long understood marriage as entailing the gift of one's body to the spouse in perpetuity,[4] so that the churches, and Christian states, until very recently failed to recognize the possibility of rape in marriage. Still today not all the United States recognize marital rape, the largest single type of violent rape, and until 15 years ago no state recognized it. In Judaism also there has been some blindness to illegitimate exercise of men's power in families. The Mosaic law's list of relationships covered by the incest ban includes bans on a man uncovering the nakedness of his mother, his sisters, his granddaughters, his father's wives, his daughters-in-law, his aunts by blood and marriage, his sisters-in-law, but not his own daughters (Lev.18: 7–16), though father-daughter incest is one of the most prevalent, most violent, and most damaging type of adult-child incest ([44], pp. 231–232). The women covered by the ban were understood as the property of another man, while his daughters were his. In the Decalogue, men are forbidden to covet not only their neighbor's house, ox, ass, manservant or maidservant, but also his wife – another piece of his property (Ex. 20:17). In the Mosaic law, if a man rapes a betrothed virgin, he is sentenced to death; if he rapes an unbetrothed virgin, he must pay her father the bride price and marry her (Deut. 22:25–29). The injury done is understood to be to the victim's father or betrothed, not to her; she can be married off to her rapist.

EFFECTS ON WOMEN

Just as there are some disadvantages to men in the power given to men in the patriarchal family, there are some limited benefits which accrue to women despite their subordination and restriction to home and motherhood. Chief among the positive benefits are a capacity for and interest in relationship with others which frequently produces among women intimate friendship, varying degrees of solidarity, and nurturing bonds to children. One must be careful not to understand these as inherent in women; just as some men resist male socialization to develop these qualities, so some women do not develop these qualities. But by comparison to men as a group, women are inclined to be much more relational and to develop skills for nurturing intimacy. For example, studies of conversation among male groups and female groups demonstrate that men are more likely to interrupt one another, discuss impersonal subjects and events rather than feelings or relationships, and to compete with one another for the speaker role. In contrast, women's groups demonstrate attempts to include women who have been silent, to share feelings and discuss relationships, and seem to lack the dominance hierarchy so central to men's groups ([2], pp. 292–299).

Yet for all the importance of these relational skills, the overall effects of women's role in the family have been negative. The understanding that women were created to be full time mothers and domestic workers has resulted in economic discrimination against women for millennia. Even today in the US the average female wage for full time workers is less than 2/3 that of men, though women as a group are better educated than men. The channelling of the majority of women into "women's work," sometimes called pink collar work, makes "comparable worth" campaigns difficult to implement, and so women's work continues to be underpaid. Furthermore, the majority of married women who do work today find that they, compared to their husbands, carry a "double burden" [1]. They share the burden of supporting the family, in addition to virtually the entire burden of child care and domestic work. Working wives have a workday of ten to twelve hours; they average workweeks of 76 hours ([18], p. 379 in [1]). Studies of men's labor in the home report that husbands spend from .3 to 1.3 hours per day in child care or domestic labor ([37], p. 285 in [1]). Thus in the contemporary world what might once have been an equitable division of labor – though not of power – between men and women in the patriarchal family has become blatantly inequitable as women share the burden of men's traditional role, while men fail to share the burden of women's traditional role.

The definition of women as made to be mothers within the patriarchal family contributed to an understanding of women as body, as inherently carnal, during centuries when all things material, and especially the body, were understood as morally dangerous and inferior to the spiritual soul, which characterized men. It was the possession of the soul that was understood to make humans reflect the image and likeness of God; yet Christianity tended to follow Augustine's dictum that men possessed the image and likeness of God in themselves, women only when joined to a man (De Trinitate, IV, 6). The carnality which characterized women for most of Christian history was understood to justify men's control of female bodies. This not only allowed the sexual abuse of women described above, but produced many other variations of female loss of control of their bodies. It was decided that the evil seductiveness of female flesh (to men!) should be covered and kept out of sight; women who failed to heed this decision were considered to have invited random abuse. Women's carnality was considered to be redeemed only through childbirth (I Tim. "Yet woman will be saved through childbearing, if she continues in faith and love and holiness, with modesty"). Because childbearing was women's function, until recently Christianity has forbidden women any reproductive control over their bodies: no contraception, no abortion, and no right to refuse intercourse in marriage, even though frequent childbearing endangered women's lives, and millions of women died in childbirth. Women still struggle for control over their bodies with a largely male medical establishment,[5] which, it is charged, has developed medical procedures contrary to women's interests, such as widespread caesarean sections, induced births, unnecessarily radical hysterectomies and mastectomies, not to mention the development of the horizontal birthing position which extends labor time [48], [16].

But perhaps the most negative effects of women's role in the family have been on women's personalities and identities. Many women in the West feel a lack of individual identity. Longer lives and shorter periods of child bearing and rearing have left many women feeling as if they have drifted through their lives fulfilling assigned roles without any sense of themselves as individuals. Universities, churches, and volunteer organizations benefit from attempts of middle class women to fill this void; marriages are strained as women seek to expand their interests and test their capacities in new ways. There is frequent feminist discussion of women as spiders sitting on a web and unable to distinguish themselves from their webs: they are daughters, sisters, wives and mothers, but they fear that apart from these relations they do not exist, and are therefore dependent [27]. The contemporary understanding of the human person as autonomous contributes to their dissatisfaction [27], [47]. This is the reverse of the situation of many men, who have constructed the identity which women feel they lack but are constrained from the relationality which women feel defines them.

Passivity is often described as characteristic of women. One element of passivity under increasing attention is fear. There is little doubt that most women live in fear, to greater or lesser degrees, of male anger and aggression. A cursory review of the statistics on domestic battering and sexual abuse makes this understandable. But we are coming to understand that the 20, or 30 or 40 percent of women who personally suffer the various types of male abuse are not the only women who live in fear. The knowledge that one's mother, sister, friend or neighbor (not to mention many anonymous women in the daily news) has been beaten, raped, or molested is sufficient to provoke fear of men's anger and aggression. Even fathers teach girls to fear violence, or at least coercion, from males; many end their warnings with "I know; I was once a young man." In fact, parents of both sexes teach women from the time that they are little girls that men are to be feared at some level. "Don't talk to strange men, don't accept rides, be careful to travel in groups, always carry money for a cab or phone." The very measures we teach our girl children to make them safe, alert young women that there is danger out there, and it is male. Mothers often counsel children to avoid or to appease Daddy when he is angry (whether the anger is appropriate or not), thus teaching children that Daddy's anger is potentially dangerous. Fear has a debilitating effect on persons; the greater the level of fear, the more energy must go to coping with the fear, and less is left for other activities. A great deal of many women's energy gets used coping with fear and attempting to avoid, deflect, or distract men from men's anger and violence, not to mention coping with the familial after-effects of anger and violence.

Is it any wonder that religious women are increasingly admitting to problems with relating to a male God we call Father? It seems clear that the Judeo-Christian tradition, so long as it continues to identify God as male and father, and so long as women's experience of males is tinged with fear of anger and violence, will have great difficulty in overcoming traditional understandings of God as strict judge, swift to anger and terrible in his wrath.

But explorations of anger have expanded from considerations of the effects of men's anger and violence to deal with the phenomenon of anger in women. Many writers increasingly point to anger as both having beneficial effects, and as being absent in women ([17]; [34], Chs. 6, 12). Many encourage women to experience their anger, rather than repress it, with the understanding that anger arises from injury, from a slight to their personhood or that of someone they care about. If the anger is not expressed toward the cause of the injury, but is repressed and unrecognized, it takes other more destructive forms. One common form that repressed anger at injury takes in women is self-blame. Victims of rape and child sexual abuse frequently come to accept responsibility for their own victimization, rationalizing that such atrocities do not happen to the innocent, that they must have done something to attract or set off the abuser ([13], Chs. 7, 8, 10). Victims of marital rape and child incest abuse are often so dependent both emotionally and materially on their abusers that they fear the loss of love and support if they vent their anger by accusing the abuser. Even women victims of strangers frequently fear a lack of support from relatives, friends, and social agencies – a fear which is all too often realistic – and interpret this to mean that there is no basis for their anger; that it is appropriately aimed inward.

Anger aimed inward in the form of self-blame is terribly destructive, and often begins a process of loss of respect and concern for one's self as not worthy of care and protection. The fact that many victims of sexual abuse, especially young victims, are frequently revictimized later in life, is often explained through this learned failure to care for, and therefore to protect, themselves ([44], Ch. 11).

But all injuries to women are not sexual. Parents and teachers who steer top ranking girl students into nursing schools instead of medical schools, secretarial programs instead of academic tracks, companies who routinely pass over women for promotion, personnel directors who ask women applicants what form of contraception they use and do they plan to have children, all do women injury. The anger at this injury is often repressed because women know that such behavior is not intentionally vindictive, and is often well meant. But repression of such anger does not eliminate it.

All persons yearn for power, not necessarily power over others, but personal power – the power to choose and direct one's course, to make a difference, to achieve, to earn recognition. Many women in the traditional family have been denied freedom and opportunities to fulfill their yearning for power in legitimate ways. The role ascribed to them is supportive, with little room for independence. Denied legitimate methods of exercising personal power, some women, like some in other suppressed groups, choose illegitimate ways of exercising personal power, usually through manipulation of those to whom they are bonded in the family ([36], pp. 13–20). In the nineteenth and early twentieth century, middle and upper class women frequently used illness as a means of gaining attention and controlling other family members. The nagging wife, the controlling mother, are contemporary stereotypes not true of women as a whole, but nonetheless all too often real. They result at least in part from the restriction of women to a narrow role in which they are unable to recognize or satisfy their own ambitions.

Such responses of women to the restrictions of women's role in the family are still present in our world, but as opportunities for some women open up we have become aware of another type of damage done to women by socialization for the patriarchal family, and that is fear of success ([14], pp. 238–241; [36], pp. 29–32, 119–122). Many women fear success in a variety of forms. Adult women sometimes fear job success out of concern that their husbands cannot accept wives who earn more or who have higher status positions; this fear may be well grounded, as such situations frequently figure in reasons given for divorce. Many young women deliberately fail to succeed both in school and in sports, especially when directly competing with men, for fear this will turn off boyfriends, or boys in general. Some women exhibit a more generalized fear of success as something not feminine, something they are unable to reconcile with themselves as women. Women seem to achieve best when they understand the endeavor in terms of developing or testing social skills. Women are least likely to achieve when success is achieved in direct competition; studies show that while competition raises achievement for boys, it decreases success for girls, even when the competitors are same sex ([31], pp. 247–254; [20]).

Women's repressed anger and the fear women carry of men take their toll on women's relations with men, even men they have no reason to fear, men who have not injured them. They create real tension in women who want to love and trust their husbands, but cannot entirely banish the whispered warnings of male violence and male failure to take women seriously which echo in women's heads every time their husband curses a reckless driver or complains about his secretary's incompetence.

All of this leads us to one of the greatest casualties of the patriarchal family structure: the marital relationship itself. There seems little doubt that the interpersonal intimacy which can characterize marriage is rendered infinitely less possible within the structure of patriarchal marriage. Both spouses suffer this lack of intimacy, though differently. Women seem to have greater resources for intimacy outside marriage than men, who are more dependent on wives for intimacy than wives are on them. But wives are much more likely to feel a conscious need for intimacy, and to mourn or resent low levels of intimacy in marriage. It is the power relation in patriarchal marriage which operates against intimacy. When one person has power over another – and the greater the power, the more profound the effect – trust becomes difficult. The powerless party is unlikely to fully trust the powerful party, and therefore avoids complete vulnerability. Who of us chooses to confess our fears and failures on the job to our boss? Co-workers are more likely confidants, and, especially for men, only when the co-workers are not in any way competitors. We are more likely to confide in fellow students than teachers, siblings than parents, unless the object of our confiding is to enlist the powerful one on our side against another. True marital intimacy requires trust so total that it allows us to abandon self-consciousness – to fail to consider how we appear to the other – and to merge with the other at some level. To merge with someone of superior power is dangerous, for we can disappear completely, be swallowed up by the other – which has been women's experience of the legal, political and economic consequence of marriage historically understood as "two in one flesh."

While the powerful person may have no fear of being controlled, punished, abandoned, swallowed or otherwise hurt, the role division in the traditional family leaves men also wary of marital intimacy. For the nature of power over others, which men are conditioned to see as central to masculinity, is that it is never vulnerable, never allows the lowering of the barrier of self-consciousness, lest power be lost. Fear of intimacy, especially with women, as an obstacle to identity, which may be left over from boys' early struggle for gender identity, may also contribute to men's resistance to marital intimacy, and to men's tendency to understand marital intimacy in terms of physical sharing in sex ([6], p. 364) rather than as an integration of physical with emotional intimacy.

EFFECTS ON CHILDREN

The area in which the traditional family's effects have received the least scrutiny is that of effects on children. The ideological right today bemoans the demise of the traditional patriarchal family; their argument that children are the foremost victims of its demise is widely accepted. Even liberals, who support the changes in women's role opposed by the right, often approach the issue of the family by asking how to compromise the legitimate aspirations of women with the real needs of children for the protective nurture of the traditional family. There is little systematic attempt to assess the positive and negative effects of the traditional family on children.

The most commonly posited benefit of the traditional family for children is the full time presence of mothers. In fact, this supposed benefit does not withstand much historical scrutiny, as it seems to suppose a modern, stable, two parent household with a full time female homemaker. For thousands of years women in the home worked – in the fields, in animal care, weaving, slaughtering, candle and soap making, not to mention cooking, washing and cleaning. Among the poor, married women were often employed outside the home as cooks, laundresses, in factories and mines, as seamstresses and maids, and in raising the children of the rich. It has probably been historically accurate to posit that contemporary working mothers with a 40 hour workweek spend more time interacting with their children than their ancestresses did.

We also need to look at the psychological research on the effect of female child care on young children referred to above, which leads girls and boys to different strategies for reaching gender identity – strategies which promote later deficiencies in intimacy and relationship for boys, and in individual identity for girls.

Liberal rights language is just beginning to be applied to children. It can be helpful to probe the rights of children, and the extent to which children's rights are recognized in the traditional patriarchal family. Children obviously have at least equal rights to material subsistence with adults. The role of fathers as breadwinners in the traditional family is assumed to supply the material needs of children. When fathers are adequately remunerated for work, children are usually benefitted.[6] However, the material dependence of children on fathers is often experienced by children – and often consciously presented by parents – as the basis for fathers' authority in the family, the reason that obedience is due to father: "So long as you live in my house, and I feed you, you do as I say." The dependence of the child on father's control of resources (or in female-headed households on mother's) conditions children to view reality as a realm where those with material resources call the shots. While this can stimulate in children a positive desire to aspire to economic self-sufficiency as an adult, it can also have very negative effects. It is a very real factor in children leaving home before they are prepared for independence, either as runaways or in order to marry too early. It can lead children to see the amassing of material resources as the means of being in total control of one's life (especially when fathers are not only powerful but non-expressive of feelings of failure, dependency, and anxiety), when in fact such freedom is impossible; all humans must learn to deal with dependency and the unavoidable arbitrariness of accidents, sickness, and death. This link between father's resources and father's authority can also lead children to desire control over others, to be as powerful as, or more powerful than, father. Worse, it can condition children to be deferential to those in later life who have more resources than they, as if those persons have a right to authority over them. Interestingly enough, the Judeo-Christian tradition has from its beginnings criticized attempts by the rich to use their wealth to manipulate and control the less fortunate, to deny them justice and rights, but has never recognized that the identification of the headship role with the breadwinner role of men predisposes children to resign themselves to the real power – if not the just right – of the rich to manipulate others less well endowed.

The emphasis on the obedience of children to parents, in the absence of any elaboration of the rights of children, fails to prepare children to recognize parental wrongdoing, whether towards them or others. Thus child victims of parental physical or sexual abuse, even when they consciously hate the abusive parent, frequently have no consciousness that the parental action is objectively wrong and could be condemned by, or much less stopped by, outside authorities. The difference in age and status between parents and children, not to mention the tremendous emotional and material dependence of children on parents, will always make it difficult for children to challenge parents' wrongdoing. But the model of the traditional family actually encourages the child to see the parent as the ultimate authority, incapable of wrongdoing. Even when children as adults could otherwise expect to be freed from obligations of parental obedience, the religious tradition has insisted that adult children honor and respect parents, without specifying any limits on that honor and respect (Ex. 20:12, 21:17; Mt. 19:18–19; Eph. 6:1–3). While it is one thing to demand that children provide material support for dependent elderly or disabled parents, unlimited honor and respect, as indicated in the Exodus condemnation of anyone who curses his parents, is quite another.

The terror and trauma of children trapped in physical, sexual or emotional abuse by parents cries to heaven, but because children have been silenced by our failure to instill in them a sense of their own right to freedom from abuse, only heaven hears their largely unspoken cries. We often read that many of these abused children grow up to be abusers, which is true. But we need to add to this picture the hundreds of thousands of child runaways whose chief reason for flight was abuse, and who all too often end up further exploited in our urban underworlds. We need to add to the picture the thousands of victims of parental incest who become self-destructive out of internalized anger and hatred, and commit suicide, or chose the slow death of drugs, alcohol, or the daily death of self-punishing sexual exploitation by others. We have a tendency to think that these cases, despite burgeoning statistics, are only exceptions to a general rule of parental nurture. But we have a great deal of psychoanalytic literature that describes other, more subtle but also disabling consequences on children of the power roles embedded in the traditional family.

Alice Miller, a German psychoanalyst, writes of her work with many "successful" persons whose parents have demanded and received the surrender of the child's self to meet the parents' needs. The loss of the self in such children cripples them emotionally:

> Children who fulfill their parents' conscious and unconscious wishes are "good," but if they ever refuse to do so, or express wishes of their own that go against those of their parents, they are called egoistic and inconsiderate. It usually does not occur to the parents that they might need and use the child to fulfill their own egoistic wishes. They often are convinced that they must teach their child how to behave because it is their duty to help them along the road to socialization. If a child brought up this way does not wish to lose his parents' love (and what child can risk that?) he must learn very early to share, to give, to make sacrifices and to be willing to "do without" and forego gratification – long before he is capable of true sharing or of the real willingness to "do without." Frequently such children use their own children, satisfying the needs of the suppressed child in themselves at the expense of their child, just as their parent did ([35], p. xii).

The very language we use to describe many of the reasons for having children are revealing, for they say little of respect for the dignity and identity of the child her/himself: to insure our immortality, to succeed more than I did; to be the —— I never got to be; for company and security in my old age; to see myself in another person; to have someone who will love me; to have someone I can pour out my love on; to be a real woman (man). All of these reasons are based on the needs of the parent, and on using the child to fulfill one's need. The very description of women in the traditional family as necessarily mothers – and the long millennia that barren women have been despised or pitied – encourages women, in particular, to see children as fulfilling a basic need for women.

In many ways we do see children as property, a special kind of property which fulfills a variety of intimate needs for us. We give very little thought to the process by which children become adults (and later parents themselves), to how they are to move from being dependent objects to becoming subjects with free agency. Marriage counselors tell us that many of the most common marital problems stem from relationships between parent and child that remain unresolved ([15], Ch. 3). Most persons in our society marry long before having reached adulthood in terms of seeing themselves as peers of their parents. Most remain trapped by the emotional authority they were socialized to grant parents. One problem for marital intimacy resulting from this failure to throw off the emotional authority of parents takes the form of fearing to achieve greater emotional intimacy in one's marriage than one's parents achieved. The adult child has accepted the parents as the norm which cannot be surpassed; to attempt to do so is to be disloyal. Such disloyalty implies a recognition of parental failure at intimacy, not only within the parental marriage, but also a failure to satisfy the intimacy and affection needs of the child. Such recognition is often painful for the child, a festering sore that has often been long repressed.

A failure on the part of the adult child to face parental failure in intimacy and one's own resentment toward the parent for such failure (and all parents sometimes fail their children) produces resistance to intimacy with one's spouse, often most obviously in the presence of one's parents.

Another obstacle to marital intimacy reported by marriage counselors stemming from the failure of adult children to understand themselves as peers, and not emotional dependents, of their parents, concerns identifying one's spouse with the parent of the same sex. We often marry seeking in a spouse the positive qualities we see in our opposite sex parent without the negative qualities we resent in that parent. Frequently, after marriage, we come to see in the spouse the negative qualities we most resent in the parent – whether or not the spouse really exhibits such qualities. When we have never resolved with the parent the resentments which linger from our childhood, when we have never recognized and forgiven the parent their failures with us, we often read into and resent in the spouse the qualities we resented and feared in the parent but could never openly face in the parent due to dependency and fear/reverence for parental authority. The wife who resented her father's consistent failure to listen to her, to take her seriously, may explode in anger at her husband's momentary preoccupation which causes him to miss some more or less inconsequential remark of hers, and may see a few such incidents over a period of years as evidence that he is just like her father: uncaring. The husband who resented his mother's attempts to control his every action and relationship through guilt and emotional manipulation may flare up at a wife who expresses a preference regarding some decision they face; he interprets her statement not as a preference, but as an attempt to control his decision, as his mother did.

In conclusion, the emphasis on children's obedience to parents in the Judeo-Christian tradition, and the underlying assumption that parents are natural protectors guided by the best interests of the child, have worked against any recognition of the rights of children. The absence of treatment of children's rights has failed to restrain parental abuse of children, helped prevent children from resisting parental abuse, and caused social blindness to the abuse of children. In addition, because within the tradition children are always children of their parents, subject to some degree of obedience, and never become peers, children are hindered in resolving the very real and powerful resentments against parents that they carry from childhood. These resentments are often pre-rational, deeply rooted angers (sometimes unreasonable) at the ways in which they were weaned, toilet trained, disciplined, and touched or not touched with affection, or even actually abused. Religious communities could be a source of real support for children in becoming adults if they encouraged children to see their parents as humans, not as god-like authorities, humans who are themselves wounded, who make mistakes and fail to love enough. For only if children can see their parents realistically, and forgive them their inadequacies, can they move on to the real issues of adulthood without projecting the problems of childhood on their adult relationships and situations. Such a resolution of our childhood resentments affects also our relationship with God, since the Judeo-

Christian tradition presents God as parent. If we so often project our problems with our parents on our spouses, how easy it is to project these same problems onto our divine parent.

CONCLUSION

It is clear that the patriarchal understanding of the family within western religion is not merely an issue for theologians. Religion shapes both individuals and social relationships. Many different kinds of contemporary professionals – doctors, psychotherapists, social workers, and marriage counselors, among others – as well as a variety of non-religious institutions deal on an everyday basis with with unhealthy effects of religious understandings of the family. Their concern is one source of support for the growing interest within many religious communities both in revamping the family in directions which would allow for healthier individuals and relationships among men, women and children, and in using social scientific criteria to give shape to these new directions.

The greatest obstacle to such revamping is ingrained reverence for religious tradition, which manifests itself in opposition to all changes in what is perceived as the revealed tradition. The best rebuttal to such opposition seems to be a demonstration of the internal contradictions within western religious traditions, in particular the existence within these traditions of potentially liberating perspectives which counter dominant practice regarding the family. Within Judaism, we can look to the nature of Yahweh as revealed both in divine intervention to liberate the suffering Hebrews enslaved in Egypt and in the law given to Moses, which took such pains to establish a strong community rooted in justice and concern for the weak. From this grounding, both Talmud and Mishnah elaborated further protections for women and children. In Christianity, Jesus's treatment of women, children, work, and power, together with the evidence of how the early church in Acts and the epistles implemented some of these teachings, form the basis for rethinking the family. This work has only barely begun.

Xavier University
Cincinnati, Ohio, U.S.A.

NOTES

[1] See the speeches of twentieth century Roman Catholic popes, for example, [21], [22], [23]. In Protestantism, Martin Luther himself is often pointed to as the initiator of this romanticizing of the family ([29], pp. 89, 160–161, 191) though his appreciation of the joys of the family hearth and the conjugal bed are less fulsome than in the twentieth century popes, and seem a welcome relief from the often legalistic and ascetic treatment of family in the theology of his own day.

[2] Mirra Komorovsky writes: "The need to maintain a 'manly' facade, the fear of acknowledging 'feminine' traits – all generate in the male a constant vigilance against the spontaneous expression of feelings ... Such guardedness adds stress to the ever-present external sources of tension" [28]. Jourard argues that being manly requires men to wear a kind of neuromuscular armor against expressiveness. He goes so far as to consider that the chronic stress thus generated is a possible factor in the relatively shorter lifespan of men as compared with women. But the deleterious effects of such self-control do not end with stress. A man who does not reveal himself to others is not likely to receive their confidences. It is precisely in the course of such interaction, however, that a person learns to recognize his own motivations, to label emotions, and to become sensitive to the inner world of his associates. Without an experience of psychological intimacy a person becomes deficient in self-awareness and empathy [26], [3]. Still another recent writer ([11], p. 210) alleges that men are threatened by psychological probings of feelings. What passes for confidence in their relationships with women is nothing but their need for uncritical reassurance ([28], p. 158).

[3] This range is based on two studies: ([25], p. 145) for the conservative figure of 20–30 percent, and ([44], p. 158) for the 48 percent, which is conservative in its own way: Russell's survey found that 82 percent of the victims of child incest were later victims of serious sexual assault as adults, whereas *only* 48 percent of those who had not been victims of child incest suffered serious sexual assault as adults. Thus the figure of 48 percent is not for the population as a whole, but for the population who were not victims of child incest. Figuring in the victims of child incest would raise the figure even higher.

[4] As found, for example, in 1 Cor. 7:3–4: "The husband should give to his wife her conjugal rights, and likewise the wife to the husband. For the wife does not rule over her own body, but the husband; likewise the husband does not rule over his own body, but the wife."

[5] This male medical establishment itself used Christianity in its bid to replace the midwives who preceded the male physicians: midwives were accused by the male physicians of pagan superstition because of their refusal to use the new mechanical forceps and their insistence on working with, and not against nature. As a result, thousands of midwives were burned at the stake as witches by the church [10]. The struggle over women's bodies continues [41].

[6] For example, there is continuing emphasis through one hundred years of Catholic social teaching on the just wage as one sufficient for the support of the wage earner and his family in adequate comfort and security, without the necessity of his wife or children working ([29], p. 662; [26], pp. 626–629).

BIBLIOGRAPHY

1. Andolsen, B.: 1985, 'A Woman's Work Is Never Done', in B. Andolsen et al. (eds.), *Women's Consciousness, Women's Conscience: A Reader in Feminist Ethics,* Winston, Minneapolis, pp. 3–18.
2. Aries, E.: 1977, 'Male-Female Interpersonal Styles in All Male, All Female, and Mixed Groups', in A. Sargent (ed.), *Beyond Sex Roles,* West, St. Paul, MN, pp. 292–299.
3. Balswick, J.O. and Peek, C.W.: 1971, 'The Inexpressive Male: A Tragedy of American Society', *Family Co-ordinator* 20, 263–268.
4. Bird, P.: 1974, 'Images of Women in the Old Testament', in R. Ruether (ed.), *Women and Sexism,* Simon and Schuster, New York, pp. 41–88.
5. Brown, J.C. and Bohn, C.R. (eds.): 1989, *Christianity, Patriarchy, and Abuse: A Feminist Critique,* Pilgrim, New York.
6. Bunker, B.B. and Seashore, E. W.: 1977, 'Power, Collusion, Intimacy/Sexuality, Support: Breaking the Sex-Role Stereotypes in Social and Organizational Settings', in A. Sargent (ed.), *Beyond Sex Roles,* West, St. Paul, pp. 356–370.
7. Cantor, A.: 1976, 'Jewish Women's Haggadah', in C. Christ and J. Plaskow (eds.),

Womanspirit Rising: A Feminist Reader in Religion, Harper and Row, San Francisco, pp. 185–192.
8. Chodorow, N.: 1978, *The Reproduction of Mothering: Psychoanalysis and the Sociology of Gender,* University of California Press, Berkeley.
9. Clark, E. and Richardson, H. (eds.): 1977, *Women and Religion,* Harper and Row, San Francisco.
10. Ehrenreich, B. and English, D.: 1972, *Witches, Midwives, and Nurses,* Feminist Press, Old Westbury, CN.
11. Fasteau, M.F.: 1974, 'Why Aren't We Talking?' in J. Pleck and J. Sawyer (eds.), *Men and Masculinity,* Prentice-Hall, Englewood Cliffs, NJ, pp. 19–21.
12. Fiorenza, E.S.: 1983, *In Memory of Her: A Feminist Theological Reconstruction of Christian Origins,* Crossroad, New York.
13. Fortune, M.M.: 1983, *Sexual Violence: The Unmentionable Sin,* Pilgrim, New York.
14. Frieze, I. et al.: 1974, 'Achievement and Nonachievement in Women', in I. Frieze et al. (eds.), *Women and Sex Roles,* W.W. Norton, New York, pp. 234–254.
15. Gallagher, C. et al.: 1986, *Embodied in Love: Sacramental Spirituality and Sexual Intimacy,* Crossroad, New York.
16. Gordon, L.: 1977, *Woman's Body, Woman's Right: A Social History of Birth Control in America,* Penguin, New York.
17. Harrison, B. W.: 1985, 'The Power of Anger in the Work of Love: Christian Ethics for Women and Other Strangers', in C. Robb (ed.), *Making the Connections: Essays in Feminist Social Ethics,* Beacon, Boston, pp. 3–21.
18. Hartman, H.: 1981, 'The Family as the Locus of Gender, Class and Political Study: The Example of Housework', *Signs* 6, 366–394.
19. Hauptman, J.: 1974, 'Images of Women in the Talmud', in R. Ruether (ed.), *Women and Sexism,* Simon and Schuster, New York, pp. 184–212.
20. Horner, M.S.: 1970, 'Femininity and Successful Achievement: Basic Inconsistency', in J.M. Bardwick et al., *Feminine Personality and Conflict,* Brooks-Cole, Belmont, CA, pp. 40–74.
21. John XXIII: 1960, 'Ci e gradito', *Osservatore Romano,* 8 December 60.
22. John Paul II: 1979, 'All'indirizzo', *The Pope Speaks* 24, pp. 168–174.
23. John Paul II: 1979, 'Chi troviamo', *The Pope Speaks* 24, pp. 165–167.
24. John Paul II: 1981, 'Laborem exercens', *Acta Apostolicae Sedis* 73, pp. 577–647.
25. Johnson, A.G.: 1980, 'On the Prevalence of Rape in the U.S.', *Signs* 6, 136–146.
26. Jourard, S.M.: 1971, *Self-Disclosure,* Wiley-Interscience, New York.
27. Keller, C.: 1986, *From A Broken Web: Separation, Sexism and Self,* Beacon, Boston.
28. Komorovsky, M.: 1976, *Dilemmas of Masculinity,* W.W. Norton, New York.
29. Leo XIII: 1891, 'Rerum novarum', *Acta Sanctae Sedis* 23 [William Gibbon (ed. and transl.), *Seven Great Encyclicals,* Paulist, New York, pp.1–30)].
30. Luther, M.: 1967, *Table Talk,* in T.S. Tappert (ed. and trans.), *Luther's Works,* Vol. 54, Fortress, Philadelphia.
31. Maccoby, E. and Jacklin, C.: 1974, *The Psychology of Sex Differences,* Stanford University Press, Berkeley.
32. Martin, D. : 1976, *Battered Wives,* Glide, San Francisco.
33. McLaughlin, E.C.: 1974, 'Equality of Souls, Inequality of Sexes: Women in Medieval Theology', in R. Ruether (ed.), *Women and Sexism,* Simon and Schuster, New York, pp. 213–266.
34. Milhaven, J.G.: 1989, *Good Anger,* Sheed and Ward, Kansas City.
35. Miller, A.: 1981, *The Drama of the Gifted Child* [1979 *Das Drama des begabten Kindes*], Basic, New York.
36. Miller, J.B.: 1986, *Toward a New Psychology of Women,* 2nd ed., Beacon, Boston.
37. Minge-Klevana, W.: 1980, 'Does Labor Time Decrease with Industrialization? A Survey of Time Allocation Studies', *Current Anthropology* 21 (3), 279–298.
38. Plaskow, J.: 1976, 'Bringing a Daughter into the Covenant', in C. Christ and J. Plaskow

(eds.), *Womanspirit Rising: A Feminist Reader in Religion,* Harper and Row, San Francisco, pp. 179–184.

39. Pleck, J.H.: 1981, *The Myth of Masculinity,* MIT Press, Cambridge, MA.
40. Raines, J.C. and Day-Lower, D.: 1986, *Modern Work and Human Meaning,* Westminster, Philadelphia.
41. Rothman, B.K.: 1982, *In Labor: Women and Power in the Birthplace,* W.W. Norton, New York.
42. Russell, D.: 1983, *Rape in Marriage,* Macmillan, New York.
43. Russell, D.: 1984, *Sexual Exploitation: Child Sexual Abuse and Workplace Harassment,* Sage, Beverly Hills, CA.
44. Russell, D.E.H.: 1986, *The Secret Trauma,* Basic, New York.
45. Sales, E.: 1974, 'Women's Adult Development', in I. Frieze et al. (eds.), *Women and Sex Roles,* W. W. Norton, New York, pp. 157–190.
46. Sennett, R. and Cobb, J.: 1972, *The Hidden Injuries of Class,* Vintage, New York.
47. Smith, R.: 1985, 'Feminism and the Moral Subject', in B. Andolsen et al. (eds.), *Women's Consciousness, Women's Conscience: A Reader in Feminist Ethics,* Winston, Minneapolis, pp. 235–250.
48. Wertz, R. and Wertz, D.: 1979, *Lying-In: A History of Childbirth in America,* Schocken, New York.

SECTION II

THE EXPERIENCE OF CHURCHGOERS

WILLIAM STACKHOUSE

EXPANDING THE RANGE OF CLINICAL CONCERNS IN SEXUAL DEVELOPMENT

The core of this essay will be the presentation of sexuality-related findings from an internal program development study conducted by the United Church of Christ (UCC) in the mid 1980s. The major findings of this study have not to date been published elsewhere. Why do I present a set of parochial findings in the context of a book such as this? First, because I hope that the presentations of these findings will help illustrate the practical, clinical, and ethical utility of a broad definition of sexuality and a nonpathological perspective about sexuality-related experience. In this connection, I also hope that this information might serve to promote happier and healthier sexuality development in people. Second, I believe that readers may find the format of the study and its findings useful in the evolution of their perspective about what makes up aspects of sexual development. Third, the findings presented may be useful in generating extrapolations about the experiences of members of other religious bodies and how they might present their experiences and concerns about sexuality. Lastly, to my knowledge this is the only study of its kind and it is unlikely that it will be duplicated in other denominations. That it was done at all is to some extent an artifact of the history and circumstances within the UCC at the time it was done. This text seemed an appropriate place to offer a brief account of some aspects of the findings of this study.

AN APOLOGY

I believe that in our pluralistic American society it has become increasingly important to take an interfaith approach to religious matters. I often encourage secular professionals, particularly health care workers, to include religion in their work. And when I do so, I encourage them not to exclude Judaism, Islam, Eastern religions, New Age Spirituality, and the perspectives of those who may have grown up without formal religious involvement but who may have significant "religious influences" on their lives. The major body of this essay will be about the experiences of members of the United Church of Christ. My apologies. I hope that the findings presented and the perspectives embodied in this essay will be relevant to a variety of religious outlooks.

SOME ASSUMPTIONS: A BROAD DEFINITION OF SEXUALITY

It is important to make a distinction between understandings of *sex* and

Ronald M. Green (ed.), Religion and Sexual Health, 121–134.

sexuality. Sex usually refers to either genital sexual acts or a person's gender [1], [2]. Sexuality is a term suggesting multidimensionality. It is a basic dimension of who we are, and has spiritual, intellectual, cultural, and emotional dimensions. It permeates and affects all our feelings, thoughts, and actions [1], [2].

When I consider an individual's "Sexual Health" it is the life-long process of experiencing and understanding sexuality that comes to mind. This process begins with the experience and thinking of children. In their study of *Children's Sexual Thinking* the Goldmans describe the term "sexual" in the following way:

> The term encompasses the identity and roles of mothers and fathers, reflecting differences and similarities of men and women generally; the child's own identity and sex preferences as a boy or girl; the family, and marriage in particular as a sexual, not only social, institution; the growing and changing physical differences between boys and girls; the origin of babies and how their sex is determined, including their conception, gestation, and birth. The wider meaning also includes what the growing child experiences and observes about 'not having babies,' sex education and sources of sex information, and the social convention of clothing the sexual organs. In addition, the vocabulary used to describe sexual organs, behaviour, experience and matters associated with sexuality is included ([3], p.5).

In their description of "sexual unfolding," the process of late adolescent sexual development, the Sarrels ([4], pp. 19f.) list the following:

1. an evolving sense of the body – toward a body image that is gender specific and fairly free of distortion (in particular about the genitals);
2. the ability to overcome or modulate guilt, shame, and childhood inhibitions associated with sexual thoughts and behavior;
3. a gradual loosening of libidinal ties to parents and siblings;
4. recognizing what is erotically pleasing and displeasing;
5. the absence of conflict and confusion about sexual orientation;
6. an increasingly satisfying and rich sexual life, free of sexual dysfunction or compulsion – (for the majority, but not for everyone, this would include a satisfying auto-eroticism);
7. a growing awareness of being a sexual person and of the place and value of sex in one's life (including options for celibacy);
8. the ability to be responsible about oneself, one's partner and society, e.g. using contraception and not using sex as exploitation of another;
9. a gradually increasing ability to experience eroticism as one aspect of intimacy with another person (not all eroticism occurs, then, in an intimate relationship, but that this fusion of sex and love is possible).

It is worth noting that many of these aspects of "unfolding" outlined by the Sarrels might be aspects of lifelong sexual development, of "sexual health." A broad definition of sexuality implies a lifelong developmental process of experience, learning, reflection, and incorporation.

A NON-PATHOLOGICAL PERSPECTIVE

When considering the "Clinical" aspects of "Religion and Sexual Health" from an outlook based on a broad definition of sexuality it becomes important to maintain a "non-pathological" perspective. Many sexuality-related experiences are unexceptional, ordinary, normal in that they are an expected part of growing up and developing relationships with others in society. They may, for example, be characteristic of a certain age, such as puberty, or a representative element of growing up with certain cultural traditions. But to the individual *any* sexuality-related experience or concern may be felt as weighty, serious, even grave. The pathos, the poignancy of an experience, is best characterized and weighted by the individual experiencing it, especially in the realm of sexuality.

PARTNERSHIP IN PROMOTING HAPPY AND HEALTHY SEXUAL DEVELOPMENT

Promoting a happy and healthy sexual development which is based on a broad definition of sexuality and a non-pathological view of sexual experience demands a partnership of caring persons. Important members of this multidimensional partnership would include families, neighbors and friends, religious leaders, physicians and health care workers, and teachers. All these persons have opportunities to assist in sexual development through their care and support, by providing information and guidance, and being role models. Among medical providers, imagine the impact if OB/GYN physicians assisted parents with the range of their prenatal concerns about the sexual development of the child on the way, if the pediatrician assisted parents and children with aspects of continuing sexual development, if the school nurse helped too, and, finally, if the G.P., internist, or specialist working with adults and their health proactively assisted with the range of sexuality-related concerns that might be present.

THE "ASK THE CHURCHES" STUDY

In 1984 the United Church Board for Homeland Ministries, the domestic program agency of the United Church of Christ, embarked upon a three year project aimed at human sexuality program development. The centerpiece of this project was a program development needs assessment study called "Ask the Churches About Faith and Sexuality: A UCC Survey of Needs for Program Development."[1]

A questionnaire was distributed which asked UCC members about their human sexuality-related needs past and present; the availability and adequacy of sources of help in meeting those needs; their beliefs concerning the role of the church in this area; specific recommendations regarding helpful programs and resources; and further information regarding personal experience and desires for future assistance. Fourteen thousand of these surveys were distributed to church members in eleven metropolitan areas across the nation. Twenty-eight hundred surveys were returned, yielding a response rate of 20 percent. Given the length of the survey, the sensitive nature of many of the questions, and the voluntary direct mail return, this was a very adequate response.

The sample was representative of United Church of Christ members. It was 66 percent female and 34 percent male. The age distribution had a wide range, among an all adult sample. The majority of respondents were 50 years or older. Over 90 percent of the respondents were white, 7 percent were black, and 2 percent other ethnic groups. It was a highly educated group: over 60 percent had college degrees, compared to 30 percent among the general UCC membership.

THE PROBLEMS THAT PEOPLE FACE

To learn about the range of sexuality-related experiences that people in the UCC were facing, members were asked to consider thirty-four sexuality-related experiences. These ranged from common and relatively non-threatening matters such as concerns related to answering children's questions about sex to more sensitive conflicts arising from experiences such as extramarital sexual relations or being a victim of incest.

For each sexuality-related experience respondents were asked the following set of questions:

A. Have you ever had the experience?
B. Did you need help dealing with the experience?
C. If you got help, was it adequate?
D. If the help was inadequate, was it because –
   1. No help was available, or
   2. I needed help but didn't seek it, or
   3. The help I got wasn't adequate.

Members were also asked the same set of questions regarding their experience assisting others with the same list of sexuality-related concerns. "Needing help" for these questions meant needing help in order to help others with the concern. In addition, they were asked where they would go for help if confronted with one of these thirty-four concerns in the future, whether out of their own personal need or in order to prepare them to help others better. They were asked if they would anticipate seeking help from church sources only, community sources only, or both.

Most people acknowledged that they had many such experiences, although many also said that they did not need help dealing with them. In answering this way it is not clear whether they assumed that help meant professional help or if they included talking things over with relatives or friends as help.

The most common experiences, reported by sixty percent or more of the sample, included concerns about whether or not or when to engage in a sexual relationship (the most common experience for both men and women), loneliness and a desire for physical closeness, questions about physical maturation, children's questions about sex, negative feelings about one's body, questions about contraception, questions about marital expectations, and conflicts around sex roles becoming more equal.

About fifty percent of the sample reported experiencing conflicts about masturbation, premarital sex, sexual relations after childbirth, concerns about a child's sexual behaviors, and sex discrimination. From thirty to fifty percent reported experiences with physical problems related to sex, questions about the effects of aging on sexual expression, conflicts about friendship with a homosexual person, extramarital affairs, and cohabitation.

From ten to thirty percent reported problems with sexual harassment, guilt over secret sexual behaviors, questions about AIDS (remember, the survey was taken in 1985), infertility, the effect of a disabling condition on sexual expression, abortion, and sexuality and substance abuse. About ten percent reported problems with sexual abuse, working through a bisexual orientation, or discrimination because they were or were thought to be homosexual. Five percent or less reported problems related to suffering physical abuse, rape, incest, or having had difficulties following a mastectomy.

Reported instances of experience with sexual abuse, rape, and incest are low when this sample of church members is compared to the figures for studies of the general population. But the sample nevertheless shows serious numbers of persons effected by wide-ranging concerns and experiences. When a religious leader or clergy person considers that perhaps five or more of all women participating in a worship service has experienced rape, this should have a profound impact on his/her ministry with these issues. Likewise, for the physician to consider that among his/her ordinary patients there are likely to be the range of experiences reported above these findings should inform his/her approach to the care of these patients.

How do the figures reported above break down among various subgroups? In general differences were small. Incest and rape were more likely to have occurred to women. Men, who made up a third of the sample, represented half of the self-reported abusers. Older persons were less likely to report problems than those who were younger. The problems occurred in all sizes of churches and communities across the nation.

Fewer members reported helping others deal with most of these experiences than having the experiences themselves. Among clergy surveyed, however, percentages of helping others with such problems are much higher. The percentages of clergy who report having helped others with such problems is an important alternative way to estimate the extent of these problems among members of churches.

That members had a particular experience didn't necessarily mean that a problem exists that the church or the clergy person needs to respond. Perhaps members cope quite well, either on their own or with help. The results suggested, however, that many wanted additional help. More than a quarter of the respondents reported unmet needs related to loneliness, a need for physical closeness, and questions of whether or not or when to engage in a sexual relationship. A quarter of the women reported unmet needs related to premarital sex, physical or psychological sex problems, and marital expectations. The reason for most of these unmet needs is that people wanted help but didn't seek it.

Some percentages of unmet needs, although quite low, reflected the low rates of occurrence of the problems reported. For example, although only 3.5 percent of the women reported unmet needs regarding rape, this represents the majority of the women who reported having been raped, five percent. Similarly for incest, physical abuse, and sexual abuse most women who reported the problem did not receive adequate help.

How did the respondents view their potential future utilization of sources of help? When considering the whole range of sexuality-related concerns the following was found: a few would seek help from church sources only – about 5 percent; a few more would seek help from community sources – only, about 10 percent; but the majority would seek help from both their church and community sources – from thirty to sixty percent. Even for physical or psychological sex problems, where one might imagine that people would turn primarily to community professionals such as physicians or psychologists, most would turn to the church as well. Respondents in this study seem to be saying that their experience with these problems is multidimensional and that the assistance they want must come from both religious sources and the secular professions.

SIX AREAS OF NEED, SOME ORDINARY, SOME GRAVE, ALL POIGNANT

Through hand written narratives almost five hundred UCC members shared information about their personal human sexuality-related experiences. Their stories were in response to the following request: "Many persons want or need to share difficult and private aspects of their life with other persons. Please briefly tell a story which you have shared, or have never told anyone but would like to tell, about a sexuality-related concern from your life."

In reviewing the narrative responses they tended to fall into six post hoc categories. These were:

Adolescent and young adult sexuality-related experiences;
Experiences concerning homosexuality;
Experiences with extramarital relationships;
Marital problems related to sexual dysfunction and desire;
Problem pregnancy experiences;
Experiences with sexual abuse, rape, and incest.

Descriptions of peoples' experiences in these areas follows.

Adolescent and Young Adult Sexuality-Related Experiences

Fifty members between the ages of twelve and sixty-five wrote about concerns related to adolescence and/or young adulthood. Almost two-thirds of this group were within the ages of 26 and 65. Thus many were addressing problems that were left unresolved at least a decade or longer ago, and for some, more than forty years ago.

Most of the experiences reported related to the question of when to have a sexual experience or relationship as a single young person. They spoke of the "trauma" of sexual restraint, the "frustration" and "stress" of having to say "no" to sexual experience. For some who had been in an intimate relationship, even engaged to be married, the decision of whether or not to or when to have sexual intercourse was a source of deep internal conflict between feelings and values.

Most who were not involved in a long term relationship spoke of feeling "alone" or isolated from others in their process of decision making about sexual activity. They wrote about lack of access to information, lack of discussion, practical advice, and support. Most also spoke of physical loneliness, of the need to be held and to be close to someone. One single young adult wrote: "Intercourse and sexual gratification *along with tenderness and warmth* are important to me; ... life 'without' is frustrating, anxiety ridden, despairing, and angry. It's the sexual sharing with someone else I need. Somewhere along the line someone could have empathized with me rather than shoving the issue under the rug!"

Some of those who were younger reported feeling pressure from peers to have sexual intercourse while their feelings and values were unclear and unresolved. For most of these there was difficulty discussing their concerns with friends. Although a few reported having shared feelings with friends who felt similarly, most reported a lack of "honest sharing" and communication within their peer group. Some reported "fear of not being normal," of being "crazy," or of wondering "what is natural?"

A number of respondents told of emotionally "painful" experiences of engaging in casual sexual intercourse which resulted in feelings of hurt and lack of self esteem or unwanted pregnancy. One respondent described the experience of adolescence this way: "It was me against the world ... (I needed) someone understanding and not pushy but willing to help me work it out."

As a result of these feelings and experiences, respondents commonly identified one or more of the following kinds of help which could have assisted them with regard to their sexual decision making during adolescence: the need for parental guidance, readiness by adults to nurture children regarding sexual ethics, information about physical development, information about sexual experience, and openness to discuss issues with adolescents.

In the realm of equipping young people in the development of their sexual ethics, the following were stated as important factors: dealing with the reality of their life situation and needs, integrating their "natural" positive feelings, providing assistance with ways of dealing with peer pressure and loneliness, and offering some sanction for the need for closeness and physical intimacy in a human context of commitment and trust.

The need for honest supportive attitudes among peers was identified as an important concern. Respondents hope for relationships among friends and lovers that permit closeness without demanding sexual activity; that allow for experimentation and risk taking; that promote mutual respect and sharing of knowledge; and that provide opportunities for growth, and encourage discussion of feelings and needs.

Respondents from each generational grouping identified the need for access to information concerning birth control and protection for sexually transmitted disease. All especially desired knowledge of physical development and sexual function as preparation for both puberty and adulthood.

The need for tenderness and warmth, to be held, loved and needed was a common appeal from these respondents. It was the underlying statement of need from those speaking from the perspective of their young single years.

Experiences Concerning Homosexuality

Sixty persons presented concerns regarding homosexuality-related experiences. These persons told either of experiences or feelings regarding personal homosexual identity, unresolved sexual identity, or homosexual friends or family members.

Those who identified themselves as gay or lesbian wrote primarily about coming to terms with the risks of social and interpersonal rejection from the heterosexual majority. Words commonly used to describe this included fear of being "condemned," "betrayed," "rejected," or "shamed." These feelings had been internalized for some. Homosexual feelings for some contributed to difficulties with sex role identity, what sort of male or female to be.

For those who spoke of fantasies, feelings, or experiences which lead them to consider that they may be bisexual, there was much struggle with lifestyle decisions. Bisexual feelings lead to complicated heterosexual marital dynamics and difficulties with long term relational commitment.

A major dilemma for some gay/lesbian respondents was the perceived threat to vocational standing if one is "out," that is, public about a homosexual identity. The issue of voluntary personal disclosure presented a constant ongoing challenge.

A majority of respondents indicated that they had either "coped" with their gay/lesbian identity and its ramifications independently or with some assistance from friends or family. But most found friends, family, and professionals inadequately equipped to help them. They did not feel that the responses of others contained the requisite objectivity to be genuinely helpful.

Respondents who were parents and friends of gay and lesbian individuals told of their struggles coming to terms with the identity of their loved one. Three major concerns were identified by parents in this category: 1) the need and desire for communication with their gay son or lesbian daughter; 2) the need for peer support, conversation, and assistance from other parents of gays; and, 3) help in finding ways to communicate with and educate other family members and family friends.

Many parents, whether intentionally or unknowingly, serve as mediators and interpreters with their community and church concerning gay/lesbian identity and life-style. Both gay/lesbian and heterosexual family members spoke of the troubling and harmful results of the "conspiracy of silence" about homosexuality.

The reality of AIDS had not yet pervaded the life of the UCC at the time of this study (1985). In the few instances where it was brought up there was concern and fear for the potential for double discrimination/rejection because of both gayness and AIDS.

A very interesting dimension of sexuality presented by these respondents was the struggle with close friendship, trust, and affection across sexual orientation. Among gays and lesbians open about their sexual orientation and heterosexual friends of these "out" gays and lesbians, the sharing about and comparison of life's joys and struggles across opposite sexual orientations seemed to enhance self-understanding for both persons.

Experiences with Extramarital Relationships

Almost a hundred respondents told of experiences with sexual affairs while married or with a married person. Almost three fifths of these were women members telling about their extramarital affairs, one fifth were male members telling about their extramarital affairs, one fifth were women reporting about the infidelity of their spouse, and a few were men reporting on the infidelity of their spouse. Three fifths of these respondents were between the ages of 36 and 50, one fifth were 26–35, and one fifth were 51–65.

Of those who wrote about a personal experience with an extramarital affair, one third told of a marital problem or problems. Many reasons were given for having or seeking an affair. Some foresaw the eventual dissolution of their marriage. Those whose marriages had dissolved wrote that it had been "mediocre," "emotionally abusive," "sterile," or "disastrous." A significant number spoke of a spouse's medical problems which caused them to seek intimacy or comfort elsewhere, with varying results. Some reported that an affair resulted from a love relationship with a close friend while still in love with their spouse.

Respondents spoke of internal personal conflicts or problems which contributed to involvement in an affair. These issues included the lack of personal security or self acceptance, including the lack of a clear sexual sense of self. Some women identified conflict over their family role or "not being treated as an equal" as an impetus for extramarital involvement. Both men and women spoke of feeling "used," "ignored," or "unappreciated." Substance abuse was also identified as a factor in seeking an affair.

In addition to identifying problems contributing to extramarital affairs, many wrote of needs that led them toward or placed them in an extramarital relationship. While similar to the earlier descriptions of problems with marriages, these "needs" offer more specificity. Persons indicated they needed: 1) "more understanding" 2) "warmth," "kindness," closeness, "to be cared for"; 3) "for us to grow in the same way," "to be fulfilled in my marriage" 4) "self-acceptance," "self-appreciation," "self-esteem"; and, 5) "differentiation between 'filial friendship' and 'physical relationships.'"

Most of these respondents did not seek help in dealing with aspects of these extramarital affairs. If they did not seek help, it wasn't because they didn't need it. As one said, "I might have benefited from a long term comfortable access to a knowledgeable counselor regarding many aspects of my relationship. It took many years of accumulated difficulties to bring me to seek assistance on them." Several felt there was "no place to go," and many who sought help found the counselors they saw not adequately prepared or responsive.

Persons had several perspectives on the help they might want if they needed it additionally in the future. In general they indicated they would want support that would be "non-condemning," "comforting," and "empathetic." They indicated they would want mature open assistance from numerous sources: friends, clergy, and professionals such as doctors and counselors.

Marital Problems Related to Sexual Dysfunction and Desire

Problems of sexual dysfunction, of diminished sexual desire, and/or lack of sexual fulfillment as a marital problem were presented in forty stories. Over half of these were women reporting on their own problems. Almost thirty percent were reports from males who reported a spousal problem. Therefore, over eighty percent of those presented were female problems; less than twenty percent were females reporting on spousal problems. In this group of respondents persons from age 26 to over 66 were included. There were a variety of problems at all ages.

Problems with marital sexual relations identified as related to childbirth and parenthood were reported among the younger respondents. These persons wrote about the pressures of two career marriages combined with parenting, resulting in: conflict over divisions of labor; "tiredness" and the inability to achieve "orgasm as often"; the difficulty of adjusting to sexual feelings in relation to these new roles; and the difficulty in finding time for intimacy. Sex after childbirth was called a "trauma" for which many were unprepared.

About fifteen percent of these respondents reported problems related to lack of preparedness for marital sexuality. A large number of these indicated they were virgins at marriage, in many cases both partners. Problems in this area included: lack of premarital sexual education; intimacy inhibited by negative feelings about sex; and, the bifurcation of feelings of love and sexual desire. These concerns resulted in lack of fulfillment, routine sexual embarrassment, and in some cases continued virginity.

About fifteen percent identified lack of orgasm or "impotency" as an unresolved problem. In addition, almost thirty percent wrote about decreased or absent desire for sexual experience. Those who wrote about lack of desire linked it with depression, life changes, or other unresolved personal problems. For some, decreased sexual relations were the result of severe mental or emotional problems of a spouse.

A significant number in this category identified overall lack of communication with a spouse and specifically lack of verbal intimacy and/or reluctance or inability to speak about sex as reasons for decreased desire and/or decreased sexual relations.

Thirty percent of the respondents in this group wrote about sexual problems in marriage as a result of distrust of the spouse and/or conflict over issues of expectations, morality, or attitude. Conflict arose because of: an absence of mutual respect; not feeling valued by one's spouse; disagreement over the purpose of sexual experience and expression; and, offense at certain desired sexual behaviors. In the most extreme cases, sex was defined as a "weapon" in a bad relationship.

The following factors were described as key factors important to the success of those who resolved their specifically sexual problems within their marriage: "great effort"; a sense of mutual support; vigilant communication; and a sense of humor.

All persons dealing with these problems reported some deficiencies in the adequacy of the responses of professionals they consulted – physicians, counselors, and clergy. All reported a desire for a higher level of expertise and sensitivity.

Problem Pregnancy Experiences

More than thirty persons wrote about their experience in facing or helping another with a problem pregnancy. These included both men and women. The overwhelming number of these persons made the decision to have an abortion. A few wrote about their choice to complete the pregnancy and care for the child or give the child up for adoption.

The majority of men and women who wrote about the decision of whether to continue or terminate a pregnancy termed the dilemma as "agonizing," "private and personal," "torturing," and "lonely." All, whatever their choice, recognized the gravity of their decision and its existential nature, and noted that the decision was made in the midst of life in a moment defined by complex circumstances. The following were noted as important elements of the circumstances for many facing the decision: limited life options at the time of the pregnancy; severe financial restrictions; problems with a spouse or boyfriend which threatened long term security; social taboos about pregnancy and maternity outside marriage; personal psychological problems; and the lack of access to safe legal abortion.

All those who wrote about these decisions continued to rethink and reevaluate their decision, no matter how many years ago it was made. The decision seemed to have had a profound impact on their attitudes toward abortion and parenthood, and their relationship with their church.

Those who reported speaking about their experience with others or seeking help emphasized the importance of discretion and confidentiality. The overwhelming majority indicated that they would have benefited from additional counsel. Those who were married indicated the importance of spousal support and understanding both during the experience and for long afterward, for years following. Some felt emotionally hurt by either the insensitive treatment of physicians and nurses, the lack of a support person with them during an abortion, or the abuse of anti-abortion picketers. Many mentioned that they would have appreciated counsel from clergy.

An overall comparison of the stories told by those who continue to be "tormented" by their decision and those who now "accept" their decision suggests that supportive counsel was the determining factor. Those who made an "informed" decision, who not only now accept but also affirm their decision, had the involvement of at least one or more personal and/or professional counselors.

Further evaluation suggests that those who face problem pregnancy, both male and female, can benefit from the involvement of supportive others. This involvement should not only be during the primary crisis, when one may feel vulnerable and threatened by lack of self acceptance or social acceptance, but also when past decisions are reviewed in light of significant life changes. Like all crisis experience, difficult decisions regarding problem pregnancy are not limited to a one-time impact in a person's life. The experience and the decision continues to inform one's understanding of self and relationships long after the experience.

Experiences with Sexual Abuse, Rape, and Incest

Fifty-eight persons shared information about dealing with sexual abuse, rape, and incest. Eighty-six percent of these stories are from persons who were the victims of rape, abuse, or incest. Fewer than 9 percent are reports of a friend's or relative's experience of victimization and five percent identified themselves as abusers. Five percent of the victims were male and 95 percent female.

Among these persons, 43 percent reported about experiences with incest, incestuous molestation, or attempted incest. Around 16 percent wrote about experience with rape. About 15 percent wrote about abuse or attempted abuse but did not specifically describe the abuse or their relationship to the abuser.

Respondents ranged from ages 19 to 66. When did the abuse occur in their lives? All those who experienced marital rape were adults during the time of occurrence. About half those who had experienced rape were young adults (between 19–25) at the time; the other half were teenagers when raped. Fewer than 8 percent of those experiencing incest, incest molestation, or attempted incest were of young adult age at the time of occurrence. Ninety-two percent were 16 years of age or younger and 56 percent were 1–12 years of age when they were victimized.

Who were the abusers? In this group of respondents, all the victims knew the abuser or rapist. All those who reported having been raped named a boyfriend, date, or acquaintance. Those who experienced incest or attempted incest named fathers most often and also brothers, brothers-in-law, uncles, and cousins. Those who were molested named neighbors and family friends. A few named doctors, bosses, scout masters, and clergy.

How did these victims describe their experiences? Most of the victims speak of their experience as "a nightmare," "a traumatic experience," "a scar on my life," a "damaging" experience – "the mental problems were worse than the physical." One respondent spoke for all when she said the experience "affected my whole life." Another said, "it almost destroyed me."

What were the issues which made survival and coping difficult for victims? Almost unanimously, they reported feelings of "shame," "guilt," or "self-doubt." Dealing with the anger and hatred felt toward their abusers was very difficult for all the victims. Some mentioned a resulting "poor sexual image" and "self deprecation."

Disclosure of the experience of rape, incest, and abuse was a most difficult problem for these victims. It was especially difficult for victims when children or as victims facing marital rape. Most did not tell anyone at the time of occurrence but waited until years later. Guilt and concern about what others would think were major barriers to their disclosure of the experience.

Two barriers to disclosure were presented by almost all who were victimized as children. These were the fear of not being believed and the fear of disclosure threatening the survival of the family. Confirming the fear of not being believed, many reported experiences of disbelief. The fear of family survival was concern that the family would "break apart" or that there would be serious conflict between the parents or with siblings.

Reflecting on the issue of disclosure, some suggested that had the community climate been different they might have sought assistance. One said, "I ... was not even able to think of it until all the publicity about the commonness of this behavior surfaced through the media, talk shows, books ..."

Many of these people sought help later in life, five to thirty or more years after the experience. Among these, many report friends, family members, and spouses whose responses were of "shock," disbelief," being "stunned," and experiencing "anger, threatening violence." Many reported that counselors and therapists responded in a manner that added to their sense of self-blame.

All of these people would like to see compassionate, supportive counsel available from many different sources – family and friends, church, and trained professionals, including police, doctors, and therapists. These victims saw the church as having an important role to play in their healing process.

CONCLUSION

This essay has presented the responses of members of the United Church of Christ about thirty-four sexuality-related experiences and about six post hoc categories derived from their narrative reports of personal sexuality-related experiences. While this presentation referred to the experiences of persons in one Protestant denomination, I hope it expands our view of the range of sexuality-related experiences in an individual's sexual development. I also hope that by adopting the sort of broad understanding of sexuality and the non-pathological perspective which provided the foundation for this study, we may become more open to the range of clinical and ethical concerns people face based on their own assessment of the gravity of these concerns.

New York, New York, U.S.A.

NOTE

[1] The findings of "Ask The Churches About Faith and Sexuality: A UCC Survey of Needs for Program Development" presented herein are from a preliminary report presented to the Corporate Members and Directors of the United Church Board for Homeland Ministries, November 1, 1986. This 87 page report includes detailed methodology, additional findings to those presented herein, and detailed charts and tables. It is available from the Secretary for Human Development Programs and Concerns, The United Church Board for Homeland Ministries, 700 Prospect Ave., Cleveland, Ohio, 44115–1100.

BIBLIOGRAPHY

1. Nelson, J.B.: 1978, *Embodiment: An Approach to Sexuality and Christian Theology,* Pilgrim Press, New York/Philadelphia.
2. Carrera, M.: 1981, *Sex: The Facts, the Acts, and Your Feelings,* Crown, New York.
3. Goldman, R. and Goldman, J.: 1982, *Children's Sexual Thinking,* Routledge & Kegan Paul, London.
4. Sarrel, L.J. and Sarrel, P.M.: 1979, *Sexual Unfolding: Sexual Development and Sex Therapies in Late Adolescence,* Little Brown and Co., Boston.

SECTION III

CLINICAL ENCOUNTERS WITH RELIGION

JULIAN W. SLOWINSKI

SEXUAL ADJUSTMENT AND RELIGIOUS TRAINING: A SEX THERAPIST'S PERSPECTIVE

"Sex is a force that permeates, influences and affects every act of a person's being at every moment of existence. It is not operative in one restricted area of life (that is, simply physical intercourse) but it is at the core and center of our total life response" ([33], p. 1).

This quote from the recent writings of a Roman Catholic priest reflects a truth and a value that has many implications for a variety of audiences, including the philosopher and theologian and certainly for the clinician who treats persons concerned with sexual and religious issues.

As a clinician who has dealt with the personal journey of many people seeking to understand and effectively cope with sexual concerns, I have witnessed the struggle of those individuals who have tried to reconcile earlier religious training and experiences with their current understanding of their own sexuality. I have also observed how a person's attitudes and ability to function sexually have been influenced by religious beliefs about sexuality, regardless of his or her denomination of origin. Even in cases where the person expressed no previous religious affiliation, the cultural effects of a common Judeo-Christian heritage were often prominent in the thinking and behavioral patterns of the individual or were reflected in the sexual interaction of a couple.

I have often been struck by the frequency of superficial knowledge of religious teachings shared by the public and how much of an impact and influence these often erroneous beliefs have on people's psyches and behavior. In a sense, they are often well educated people trying to live their sexual lives guided by a grammar school understanding of their faith. Thus a therapist can no longer become surprised, for example, when treating a highly educated professional woman who recoils at the discussion of masturbation during sex therapy with the disquieting flashback of the "nuns telling us that our vaginas would burn forever in hell if we touched ourselves." As with this example, the therapist often finds that a simple cognitive understanding of the place and nature of masturbation may not bring the emotional calm intended because of the patient's long-standing discomfort with her own body and sexual feelings.

As a beginning, I would like to develop themes begun in previous essays and reflect on some of the background influences from the more than two thousand years of Judeo-Christian heritage that have shaped our response and understanding of human sexuality. How we got to be the way we are is a good starting point before exploring further the influences that, in my view, religion has, both positively and negatively, on our sexual adjustment.

Ronald M. Green (ed.), Religion and Sexual Health, 137–154.

BIBLICAL THEMES

The Bible is the usual starting place and often the final authority for formal Western religious teachings regarding sexuality and what constitutes proper behavior. From the outset it may be said that the Bible does not give absolute prescriptions regarding sex. Scholars remind us that neither biblical Hebrew or Greek has a word to express the concept of human sexuality that we use today ([8], p. 71). The observation has been made that biblical instructions on sexual behavior are culturally conditioned and have their validity limited to the context of those times. It can be said then that the Bible is not a code book on sexual ethics and regards sexuality as one aspect of life within the context of the whole community.

From the best that we can tell, the ancient Jews viewed sexuality as a natural and good part of living. They apparently did not make the body/soul distinction that the Western Christian tradition has made. The human being was an integrated whole, created by God as good and in God's image. The place of sexuality in everyday life in ancient Israel was within a patriarchal culture and within a procreational context since the survival of the nation was at stake. Mutuality of the male and female bond was emphasized as an end of marriage, much as in the later Christian tradition. Jewish laws governing sexual practices were prescriptive and social rather than interpersonally "moral." Thus "wasting the seed" had social significance for the community and its numbers.

The attitudes about sexual behavior both for ancient Jews and early Christians were strongly influenced by their continual reaction against the sexual practices of paganism. Pagan cults often included sexual practices as part of their worship, while Judaism was monotheistic and did not divinize sex. Yahweh did not have a female consort and humankind was not the product of sexual encounters of divine beings. Christians have regarded the divinity in asexual terms and to consider Christ as a sexual person causes some believers considerable discomfort. A recent example is the controversy about the film adaptation of Kazantzakis's novel, *The Last Temptation of Christ,* in which Jesus is presented as a man who is no stranger to human sexual desires.

The New Testament, like the Old Testament, does not provide us with a system of sexual behavior. While Christ's words in scripture did not directly deal with sexual behavior, his insistence on the primacy of love continues to have a strong influence on the evolution of our current ethic on sexuality. Scripture scholars remind us that New Testament references to sexuality occur within the context of the times and that to invoke them in dissimilar situations today violates the integrity of scripture [22]. The scriptural admonitions serve as a paradigm to which we can add our two thousand years of tradition and scientific and psychological data to both assist and perhaps even confuse our efforts at ethical decisions.

If we consider the teachings of St. Paul in formulating our understanding of our Western view of sexuality, we need to be aware of the fact that Paul was influenced by Stoicism, the reigning philosophy at the time. Consistent with this philosophy was a distrust of passion and desire, as indifference to things sensual was seen as virtuous. This attitude was in contrast with the Old Testament which saw no such need to discipline the senses. The eschatological viewpoint that shaped early Christian thinking lent itself to the foregoing of pleasure for the sake of the kingdom.

Unfortunately, Paul's words were later interpreted within the context of Neoplatonic dualism. This separation of the spirit and the body placed human nature in opposition to itself for centuries to come. Heroic acts of asceticism and embracing of celibacy as the preferred life style ensued. The body became despised and the spirit was exalted. This led to a difficulty for Christians in reconciling a full sexual life with a life of the spirit. The problem exists today and is a focus for theologians, ethicists, and clinicians [31], [18], [19].

EARLY CHRISTIAN AND MEDIEVAL TEACHINGS

The primacy of celibacy and virginity continued in the early Christian centuries and the Church began to legislate sexuality and marriage. Sexual intercourse became to be viewed even more as procreative in nature, and abstinence was seen as a vehicle to limit passion. It was in these early centuries that St. Augustine's writings became so influential. Augustine's own earlier contact with Manichaeanism and its view that sexual desire itself was sinful became the foundation for teachings about sexuality that still influence many churches today [21]. Augustine went so far as to say that sin is rooted in human nature and is propagated by it [8]. While procreation remained the end of marriage, spontaneous sexual desire was proof of the penalty of universal Original Sin.

The pairing of sexual intercourse with the transmission of Original Sin has profoundly influenced human beings' view of themselves. Not only did sexual desire become sinful, but human nature was corrupted by Adam's sin. Augustine's teaching transformed Christian teaching from one of freedom to choose good or evil to one of human bondage [21]. Augustine argued that semen itself is already "shackled by the bond of death," and transmits the damage incurred by sin (*De Civitate Dei,* 13, 14). Sexual arousal, apart from any action, is also sinful. One does not have to be a clinician to imagine the damage that this line of thinking can do to a person who is struggling to integrate sexual feeling and behavior in a relationship. In my own practice I have seen many examples of inhibited sexual desire that appear to be a result of a fundamental discomfort with sexuality and the mislabeling of one's sexual feelings as evil or sinful. The dilemma becomes more complicated when the subject of homosexual orientation is introduced. I will address homosexuality later in this discussion.

The centuries that followed Augustine were marked by the view that pleasure was evil in that it impaired the exercise of judgment. Today, we can easily understand how passion can interfere with reason. It was Thomas Aquinas who emphasized the supremacy of reason and the distrust of passion. For Thomas, while sex was God-given, concupiscence became the enemy of reason. Even within marriage some element of sin was connected with venereal pleasure. Thomas's thinking was influenced by the philosophy of Aristotle and he insisted on the primacy of Aristotle's notion of the Natural Law, a foundation which became the point of reference for Catholic church teachings until today.

According to the Natural Law principles, behavior can be viewed as according to or contrary to nature, as "perfect" or "imperfect." (As a point of interest, genitals were referred to by Aquinas as "dishonest parts.") Thus in a large sense, sexual behavior came to be regarded as to whether it could meet its natural end; i.e., lead to procreation. Behavior such as masturbation and birth control or questions about homosexual orientation or behavior are measured as to their standing within the Natural Law. This led to the development of moral absolutes based on traditional understandings of nature, a viewpoint which allows little freedom for change, especially when working within the traditional Catholic framework. Only recently within Protestantism have the insights of reason and experience been added to scripture and tradition as foundations for making ethical decisions. I want to emphasize that these old beliefs and systems of understanding nature are deeply ingrained in our culture, our religious institutions, and in our psyches. They can sometimes, but not always, be the source of confusion, pain, and even alienation from each other and turmoil within the self.

In summary, even with individual denominational differences aside, we have inherited a dualistic paradigm about sexuality from our religious past. This paradigm includes assumptions that considerably affect the formation of conscience of both believers and non-believers alike. The attitudes about sexuality hold a strong dualism between body and spirit, a patriarchal outlook, continued emphasis on procreation as an end of sexuality, a denial of the relationship of erotic love to divine grace, and a strict set of rules that govern sexual behavior ([32], p. 21).

In a lighter vein, sex educator Sol Gordon has summed up the collective attitude in our culture as being one of "sex is dirty, save it for someone you love." In my own experience in teaching both sex education and theology before becoming a clinical psychologist, I was continually challenged by students who wanted to know how, if sex is good, could it be a sin the night before a marriage, but a good to be shared after the wedding ceremony. These young people were amused and confused by the sacramental sanitization of sexuality provided by traditional church teachings.

Fortunately, there has been progress among theologians and ethicists in viewing the nature of human sexuality. However, I feel that their efforts at a greater, more enlightened understanding of sexuality have not filtered down to the society at large. In time, perhaps, as generations of clergy are trained with a greater appreciation of human beings' sexual nature, the general public may benefit. My concern is that, as with many issues in which institutions lag behind, the people may vote with their feet and find religious teaching irrelevant to the understanding and indeed the celebrating of one's sexuality.

NEW TRENDS

On the positive side, there is a movement towards a more holistic understanding of sexuality [33]. Yates proposes a new paradigm that views sexuality in a positive sense, is egalitarian in its treatment of men and women, views sexual behavior (either heterosexual or homosexual) as an expression that should lead towards wholeness, avoids sexual dualism, allows for the expression of a variety of loves, including the erotic union, and allows us to know God's grace through exercising our sexual nature.

Yates joins ethicist James Nelson, who has been an advocate for a more incarnational understanding of sexuality. His works have been influential in creating a sex-positive model for those who wish to get away from the tyranny of sexual dualism and the confines of the Natural Law approach [14], [17], [18], [19], [20]. I have found Nelson's works of great assistance in leading those patients who want to remain within a religious framework as they try to cope with contradictions within their sexual selves.

For some, the "ends" of sexuality still remain an issue in trying to live their sexual lives. Ethicist Timothy Sedgewick speaks in terms of the evolution of three ends to sexuality: pleasure (sex is good and natural), mutuality (love, care, joy, and regard for one another), and generativity (which includes procreation, but not exclusively) [23]. The identification of these three ends or goods of sexuality allows for a broader approach than the strictly procreational model. For example, both heterosexual and homosexual unions fit this model; relationships need not be procreational to lend generativity to the community. This is an Eriksonian notion that many will recognize and appreciate.

It may be argued that while approaches such as those held by Yates, Nelson or Sedgewick allow for a broader understanding and for the incorporation of sexuality into a more acceptable framework, they still suggest an imposition of rules of conduct on sexual behavior. Green has questioned the relevance of applying theology to sexual ethics today [5]. In part, my clinical experience has found this criticism to be true. While those committed to a religious belief system have found insight and consolation from a more holistic understanding of sexuality (this is particularly true in my work with gay men and AIDS victims), others feel constrained by the notion that there are still external impositions upon behavior that is both personal and consensual. Are we dealing with a theological evolution that has in effect sanitized sexuality but still attempts to confine and control its expression within a theological framework that may not relate to everyday experience? I realize that it is not the task of the theologian or ethicist to reduce sexual notions to those of personal fulfillment, but it is still worth asking whether they are trying to explain or justify something that from the viewpoint of sexual science and human psychology does not need justification. The dilemma of the clinician, as well as his or her task, is to deal with what is, (as culled from empirical study and the case presentation) while being careful not to introduce what should be, as determined by some traditional moral imperative.

SEXUAL DYSFUNCTION AND POSSIBLE RELIGIOUS CAUSATION

Thus far we have explored some of the underlying religious assumptions in our tradition regarding sexuality. I would like to review some notions about the role of religious training and beliefs about sexuality that are seen as clinically related to reported sexual difficulties and dysfunctions.

Tradition and myth have long and often described impotence and sexual dysfunction as a divine punishment [25]. Abimelech, king of Gerar, was rendered impotent for taking Abraham's wife, Sarah. "But God came to Abimelech in a dream by night and said to him, 'Behold thou are but a dead man, for the woman which thou hast taken; for she is a man's wife.'" Through Abraham's intercession, Abimelech's potency is restored. "At Abraham's prayer God healed Abimelech, his wife and his slave girls, so that they could have children ..." (Genesis 20).

Greek mythology also recognized childhood sexual anxiety and its psychopathological role in adult male impotence. Melampus, the physician, was able to cure the impotence of Iphiclus, son of king Phylacus, through the aid of the gods. In the Middle Ages, Thomas Aquinas wrote that the work of demons could effect procreation and copulation. A review of medieval literature on demonology and impotence has noted a close resemblance between the etiology presented by medieval theologians and modern psychodynamic theories [25].

From a contemporary perspective, the inhibiting effects that cultural and religious tradition have on freedom of sexual expression often rests on a process that results in a learned pattern of behavior that makes sexual practice difficult and unsatisfying. Echoing Sol Gordon's earlier quote in this chapter, English and Pearson present the problem in our culture:

> A child is brought into a world filled with affection and physical love; but as it grows, it is taught to resist its biologically normal responses and pull away when touched by another person. After twenty years of such training a marriage ceremony is supposed to correct all the negative responses which have been drilled into a boy or girl and in their marital relations are supposed to become unrestrained as when they were babes ([1], p. 78).

The clinician who practices sex therapy is well aware of how cultural norms present a major hurdle as both a contributing and maintaining variable in sexual difficulties. It is interesting that Margaret Mead's early work pointed out that impotence, for example, was often culturally induced and that sexual maladjustment was scarce among sexually freer cultures. Examples of the effect of religious and social inhibitions on sexual functioning are found in the works of a number of psychoanalytic writers. These authors repeatedly make the point that anxiety and fear generated from early attitudes and sexual experiences can contribute to adult sexual dysfunctions [25].

Masters and Johnson report in their original work on human sexual inadequacy that religious orthodoxy is one of the etiological factors that contributed to treatment failure in their patient population. There was 66.6% immediate failure rate in primary impotent men and a 50% failure to reverse symptoms in secondary impotent men influenced by religious orthodoxy. Sin and shame could be seen as interfering with natural biological responses ([10], pp. 211–212).

Helen Singer Kaplan has described how religious attitudes, particularly those of Orthodox Judaism and Roman Catholicism, with their prohibition of most sexual expressions outside of marriage, have contributed to the development of symptoms of sexual anxiety in her clinical sample [7]. In her earlier work on inhibited sexual desire, she also noted the connection between religious beliefs and negative consequences on sexual desire [6]. Leiblum and Rosen [9] caution not to oversimplify the relationship between orthodoxy and dysfunction, for as they point out, there are many devout people who are sexually functional and there are many sexually dysfunctional people who are not religious.

CLINICAL OBSERVATIONS

Thus far I have outlined some of the historical developments in religious attitudes towards sexuality and have mentioned how some clinicians view the influence of religion in the development of sexual dysfunctions and sexual difficulties in their patients. It is not my intention to present religion only as a negative influence on a person's sexual expression and enjoyment. I would now like to reflect on my clinical experiences with patients and the interface between their religious backgrounds and their concerns about sexual functioning.

A therapist or clinician who engages a patient in a discussion of religious beliefs is entering into a deeply personal area. A person's religious orientation is often what gives meaning to their lives, and any questioning or challenge of such fundamental assumptions about life values can be threatening. The therapist, too, approaches the therapeutic relationship with his or her own sexual and religious history and is by training expected to be aware about not letting personal values interfere with treatment. (To some extent, this level of objectivity remains an ideal.) It is the task of the therapist to confront, when necessary, inconsistencies between the religious beliefs and the expressed personal goals of that patient. This is often the case when the life experience and intentions of the patient are at variance with expressed religious teachings. These differences must be dealt with to the extent that the patient is uncomfortable with the resulting dissonance.

The most consistent and striking manifestation of the effect of religious background that I have found in my clinical practice is the continued underlying dualism between body and spirit. Despite all our modern explanations about the nature of man, we are still to a great extent saddled with the heritage of the early Fathers of our culture. By this I mean that the underlying assumption for many people, particularly those with a religious background, is that the body and its natural needs and expressions is still somewhat suspect, and matters of the spirit are elevated or better and even purer. Religious and cultural role models still exist in the example of celibate clergy, emphasis on chastity and abstinence, and the continued belief of some religious groups that the truly spiritual journey would have us transcend our bodies and, by implication, our sexual nature, and dwell in the more perfect realm of the spirit.

Such dualism, also expressed in a sexist way by placing women in a secondary role, continues to be responsible for raising generations of people, most of whom share some common religious upbringing, with certain deep seated feelings of alienation from their own sexuality. One of the most common themes I have seen in the routine taking of both a sexual history and religious history with all new patients, is the relative absence of healthy and open attitudes about sexuality in the home environment of the family of origin. If sex or one's sexual nature was discussed at all, it was usually within the context of religious proscriptions and the resulting double message, or should we say double bind, that reflects sexual dualism. Rarely have I heard that sexuality was discussed and presented within a religious context in homes as a joyous gift to be celebrated. In those cases where sexuality was presented in positive tones, it was often sanitized by religious overtones that emphasized love over eros. The erotic component was often overlooked or relegated to an unspoken and often unapproved status.

This relegation of the erotic should not surprise us, for the seeds of sexual dualism in our collective consciousness (or collective unconscious) have been there since the early Christian centuries [31]. I have seen the consequences on patients in my practice who present with a sense of guilt and discomfort about sexual feeling, sexual desire, and sexual pleasure. Here and for the rest of this essay I am referring to those people who have difficulties related to sexuality but who are for the most part not suffering from significant emotional illness or personality problems that would place them in specific diagnostic categories. I am speaking about the person who is otherwise well functioning but who is having sexual difficulties in which religious background is a significant variable. There remains a group of people for whom religiously-based problems about sexuality are part of their neurotic character which makes them especially vulnerable to sex-negative religious teachings.

THE PRACTICE OF SEX THERAPY

The practice of sex therapy is complex because it involves many variables that include not only physical functioning, but also the cognitive, emotional, behavioral, and attitudinal notions of the person. Many of these modalities and their influence and significance to the person are not always at the conscious level and it is only through exploration that influential aspects such as religious beliefs and assumptions can be elicited. In addition to specific religious prohibitions about certain sexual behaviors, one frequently finds that assumptions are made that reflect false logic and irrational leaps of faith. These are often seen under the guise of more enlightened thinking that can be more problematic than perhaps intended.

For example, I recently came across a pamphlet for Catholic parents about preschool children and sexuality. In addressing the topic of masturbation, the authors were quite accurate in presenting the current scientific evidence about the high incidence of this natural behavior in children and adolescents. However, while admitting to the high natural occurrence of masturbation, the authors go on to say that by the teen years this behavior must be stopped for it interfered with the adult ability to love another person. This leap from describing a behavior that is natural but still must be stopped is a prime example of how religion can both use and at the same time ignore the findings of modern behavioral sciences. It is also a form of theoretical eclecticism that would not stand the test in empirical science. This example about masturbation is also a statement about "bad theology" that contaminates the behavioral sciences when unquestioned assumptions are allowed to accrue the status of self evident truths.

The style of thinking described above exists throughout the interface between the behavioral scientist's view of human sexuality and that of organized religion and cultural norms. Not too long ago, it was assumed by many with a religious persuasion that the practice of birth control rendered the wife no more than a sexual object for her husband and therefore interfered with the possibility for full love between them. It seems that once the Natural End of sexuality, the procreative aspect, was interfered with, the partners were condemned to fall victim to the other side of the sexual dualism, namely to seek sexual pleasure for its own sake. There is no room in the Natural Law view for deviation from the moral absolute. Guilt induced by masturbation, artificial contraception and abortion, though certainly not equivalent practices, often is a consequence of rigid moral teachings.

When psychological aspects are finally introduced as a way of explanation or understanding sexuality, some religious institutions use outmoded theoretical notions accepted as fact and incorporate them into their argument. These notions are often early Freudian conceptualizations that are not held by current scientific investigators as a way of understanding aspects of human sexuality. In addition, a mixture of unrelated psychological theories can find their way into religious explanations of sexuality that may fit a dogmatic assumption, but may not reflect the data of behavioral science. As an example, the use of unsubstantiated theories of homosexuality by religious groups demonstrates a mixing of philosophy and science to the detriment of both disciplines. Once again, the faithful believer may be the victim when biased teachings become the framework for decision making that can induce unnecessary guilt and confusion about sexuality.

The process of the formation of conscience is cheated when available resources are not employed and the only appeal is to authority and law. When people are faced with limited choices because of their religious background, my clinical experience has been that many find their religious heritage to be irrelevant and selectively ignore institutional teachings or abandon their church membership altogether. The latter is more often the case and in my view unfortunate because many of the positive aspects of religious affiliation are lost in the rejection and at times alienation from religious institutions.

This rejection of religious teaching on sexuality and, by extension, active membership in a given denomination, is especially seen when working with gay men and lesbian women. The issue has gained greater significance with the advent of the AIDS crisis [27], [28], [29].

My recent work with gay HIV Positive and AIDS patients has made clear to me the impact that sex-negative church teaching has on gay men. Therapists often overlook the importance that a person's religious beliefs has for their sense of self worth. When faced with eventual death from AIDS, the victim can usually only respond within the psychological and spiritual framework that represents a totality of their life experience.

This life experience for many, if not most, homosexual persons usually represents membership or exposure to a church or society that condemns homosexuality. My clinical observation has been that, superficially, the gay AIDS patient may present as if he has come to grips with his homosexuality in light of his earlier religious experience. Upon further evaluation, however, this response frequently means that the AIDS victim has rationalized his sexual orientation and life-style, lived with cognitive and spiritual dissonance, and often accepted the role of an outcast. This feeling of exclusion on the basis of the teachings of one's church is also heightened by the ongoing negative social climate towards homosexuality. The frequent result is that the gay AIDS patient experiences a resurgence and intensification of his own homophobia. Religiously-induced guilt about homosexual orientation and sexual behavior can surface as a painful experience [27], [28], [29].

The religiously and socially induced guilt I am describing often has its roots in literal or traditionalistic interpretations of scripture and in a Natural Law based understanding of sexuality with its emphasis on procreation and heterosexual intercourse as the natural end of sexuality. But there has been movement among theologians and ethicists in their understanding of homosexuality vis-à-vis religious teachings. While it is true that official church teachings of various denominations are still for the most part negative in their view of homosexual behavior, there is some variation among the denominations. Nelson's works ([14], [15], [17]) have been particularly sensitive in exploring and supporting a positive acceptance of homosexuality within a Christian framework. Nelson summarizes church attitudes into four categories: rejecting-punitive; rejecting-nonpunitive; qualified acceptance; full acceptance [14], pp. 196–197). It is Nelson's position that only full acceptance of homosexual orientation can lead to the view that same-sex relationships can really express God's humanizing intentions. Any lesser view, he maintains, places the gay person in the impossible bind that accepts the person but condemns the behavior. In other words, some churches can be viewed as saying that essence of homosexual orientation is perverse by its very nature.

The issue of perversity and homosexuality is reflected in the official Roman Catholic statement that opposes homosexual behavior as "intrinsically disordered and in no case to be approved of" ([4], Declaration of Certain Questions

Concerning Sexual Ethics: 12/19/87: N. 8, Para. 4). This view is consistent with the Natural Law approach that would see homosexual orientation as contrary to nature and therefore "essentially imperfect." The natural end of sexuality would be frustrated by homosexual acts. Thus homosexuality could be placed alongside masturbation as a frustration of nature while, for example, rape and fornication might not interfere with procreation. Clearly, many Catholic theologians have moved beyond this stance, but the official teachings remain the same, or at least the average believer would so assume. Official pronouncements from Rome ([4], Letter to the Bishops of the Catholic Church on the Pastoral Care of Homosexual Persons, October, 1986), while calling for pastoral compassion, leave little expectation of a change in church teachings.

Biblical interpretations that scripture "has nothing to say on the subject of homosexuality" are viewed as "gravely erroneous" ([4], "Declaration": N. 8, Para. 4). In fact, it "is likewise essential to recognize that scriptures are not properly understood when they are interpreted in a way which contradicts the Church's living tradition. To be correct, the interpretation of scripture must be in substantial accord with that tradition" ([4], "Declaration": N. 8, Para. 5). This view has powerful implications for those seeking a more enlightened perspective. Catholic scholars are essentially being told by the Roman church to proceed with their studies but not to come up with any new ideas that are contrary to tradition. Here again, it appears that the Natural Law is presented as immutable and actions are categorized as essentially evil in themselves (in se). The finality of this statement is very often the way that gay men and women have understood their orientation, and it is often the substance for therapeutic discussion and the cause of much psychic pain over a lifetime.

What is encouraging is that Catholic theologians, including feminist reformers, are reassessing Catholic sexual ethics that focus on procreation, natural law and specific sexual acts. Kosnik and his colleagues, in a study commissioned by the Catholic Theological Society of America (but not finally approved), reflect more moderate thinking: "Wholesome human sexuality is that which fosters a creative growth towards integration. Destructive sexuality results in personal frustration and interpersonal alienation" ([16], p. 86). Kosnik develops seven values needed for sexuality to be conducive to creative growth. It must be self-liberating, other-enriching, honest, faithful, socially responsible, life serving and joyous [8]. His notions are similar to the holistic approach embraced by Yates and Nelson, but still present themselves within a value system, understandable for religion, but which as Green has pointed out [5], still places expectations and rules on sexual conduct.

While the scheme for sexuality mentioned above can apply as standards by which to measure both heterosexual and homosexual behavior, one often needs to explore more fundamental beliefs with the gay patient who seeks a more human understanding of his or her sexual orientation from within the religious tradition. Most people rely on traditional and literal understandings of scripture to form their moral opinion and prejudices about homosexuality. Modern biblical scholarship presents us with a wider view of homosexuality in the

Bible. I have found that some gay patients are helped by exposure to new ideas about biblical understandings of homosexuality. Admittedly, as mentioned in our reference to current Vatican teachings, not all churches tolerate the input of scholarship when it contradicts tradition. However, if one departs from the literalist interpretation, scripture can be viewed in terms of the context of the times. Robin Scroggs [22] emphasizes that biblical injunctions cannot be made into eternal truths independent of the historical and cultural context. He feels that homosexuality as a sexual orientation is not addressed in the Bible. Human nature was seen as one and basically heterosexual. There appears to be no understanding of sexual orientation or adult-adult same-sexed relationships as we understand them today with the benefit of advanced biological, psychological and social sciences [28].

What is essential is to understand the meaning that sexuality had for the writers and audience at the time of the writing. To be relevant, Scroggs insists, scriptural references must be consonant with the larger major theoretical and ethical judgments throughout the ages, and context today must bear a reasonable similarity to the context of the statement at the time of the writing ([22], p. 23). For Scroggs, St. Paul's judgments remain valid only against what he opposed [i.e., pederasty]. Specific references (about sexuality) cannot be generalized without violating the integrity of the scriptures. His conclusion is that the Bible does not address the issues that we are dealing with today when we consider homosexuality.

I must admit that some of Scroggs' arguments, along with those examined by John McNeil ([11], [12]) and others, are of significant help for some patients who are used to understanding their homosexual orientation within the confines of narrow scriptural interpretations. I have found that other gay men are reluctant to be convinced for they feel that they are sinners, condemned by their very nature. Gay Catholics are faced with sexual abstinence as the only way to deal with their "essentially imperfect" orientation. Consistent with sexual dualism, they are called to chastity along with everyone else who holds to the notion that sexual expression is primarily a procreative act confined to a marital relationship. In clinical practice, such a stance gets little attention or consideration from the client.

For those patients who wish to explore scripture further, they find that the biblical references to homosexuality are few (Genesis 19; Leviticus 18:23, 20:13; Romans 1:24–27; I Corinthians 6:9–10; Timothy 1:8–11). No words of Christ are reported on the subject. Specific sex acts are mentioned, but not sexual orientation in a male dominated society in which procreation and avoidance of pagan worship and sexual practices was imperative for the survival of both ancient Israel and early Christianity. New Testament writers had to contend with the influences of Greek philosophy as well as the sexual excesses ascribed to the Greek and Roman societies.

My experience has been that unless the religiously concerned homosexual patients have re-evaluated their understanding of traditional religious teachings vis-à-vis their sexual orientation, they are at risk of continuing in a spiritual dilemma. This is even more an issue for the AIDS patient and is a topic that has been dealt with by several authors [2], [24]. Therapy must be sensitive to the spiritual side of a person's nature when it is a significant issue in treatment. The spiritual is apart from the religious for it goes beyond the mere counting or not counting oneself among the membership of a denomination. Rather, as Nelson defines it, spirituality is seen as the "ways and patterns by which the person intellectually, emotionally and physically relates to that which is ultimately real and worthwhile for him or her ([20], p. 21).

While many therapists feel uncomfortable with addressing religious and spiritual issues and can inadvertently respond to the negative internalized labels that gay patients bring to treatment (such as sinner and unnatural), it is extremely important for that patient to embrace a positive attitude about the body and sexuality. Far too much energy has been spent on defending sexual orientation and assessing sexual acts as either right or wrong. I agree with Nelson when he suggests that the appropriate ethical question is what sexual behavior will serve to enhance, rather than to inhibit or damage the fuller realization of our (divinely intended) humanity ([15], p. 172). This theme has surfaced on a number of occasions in my work with HIV positive men as they struggle to make sense of their past patterns of sexual behavior.

At this point, I would like to address some of the basic issues that a therapist might see when dealing with persons for whom their religious background serves as a barrier to sexual expression and intimacy. We have already examined historical issues and basic philosophical concepts that have become part of our Western religious tradition about human beings' sexual nature. I have also suggested that while there is progress among theologians and ethicists in integrating sexuality and spirituality in a more modern framework, the average person with sexual difficulties who seeks a therapist's advice is usually the recipient of traditional teachings and often untouched by the progressive elements in religious thinking about sexuality. Unfortunately, sexual dualism still reigns for a large percentage of people and their attitudes about sexuality are more an example of sexual doctrine rather than sexual science. I might add that John Money has pointed out how sexology too can become doctrinal and perpetuate biases that reflect the culture, thus creating further cognitive dissonance [13].

PROBLEMS OF SEXUAL DESIRE

The most fundamental area where I have seen difficulties attributed to religious training is in the area of sexual desire. Becoming aware of one's sexual feelings, understanding them as natural and incorporating them into our sense of self is a primary task of our psychological development. Sexual feelings or sexual desire are genetically programmed, hormonally based phenomena that are part of being human. It is the socialization and control of them by cultural and religious norms that renders the experience, acceptance, and expression of sexual feelings to be so filled with special meaning. It is the meaning or personal, psychological value given to sexual desire that can cause so many people confusion. If our dualistic, Neoplatonic religious tradition tells us that sexual desire equals sin, or at least the presence of temptation and a concrete example of our fallen nature, then sexual desire can become a natural experience that can lead to anticipatory anxiety rather than to pleasure. What results is an internal sexual repression of sexual feeling with the consequent inhibition of sexual desire. In other words, a person can be uncomfortable with internal sexual desire, label it as wrong, and feel guilty for a genetically programmed (God given, if you wish) phenomenon. As a result, sexual expression can be repressed or the pleasurable value can be denied.

In some situations, the manifestation of this problem is in the development of sexual dysfunctions: inhibited desire; erectile failure; rapid or retarded ejaculation; female anorgasmia; vaginismus and dyspareunia (painful intercourse). Sexual desire and its consequence, sexual arousal, can in some people be the catalyst for anxiety and guilt rather than pleasure and letting go. For example, I have seen many women in my practice who complained of not being orgasmic. In some cases, it became apparent that they were shutting down their sexual response by interpreting their level of natural sexual arousal with anxiety rather than with pleasure. Their distrust and discomfort with their own pleasure also left them with further guilt when in their own mind they failed to satisfy their sexual partner.

Their experience is not too different from men, but male dysfunctions can be more dramatic when erectile and ejaculatory difficulties are involved. The impact on the relationship is also more dramatic if the sexual dysfunction is the cause for the couple's inability to conceive [26]. In these circumstances, the guilt towards the spouse compounds the original religious guilt. One can argue from a dynamic perspective that the person ends up punishing himself or herself for having sexual desires by causing the sexual failure and subsequent interpersonal disappointment. It is not my intention to present a complicated problem in simplistic terms. The development and maintenance of the kinds of sexual problems I have seen is an involved process, subject to many variables, including learned experiences and anxieties resulting from previous failures. Thus, some sexual problems become functionally autonomous and may continue even after therapeutic intervention has uncovered an underlying cause that may be rooted in religious teachings.

One of the complaints I have seen more recently is that of a lack of sexual satisfaction in persons who are not experiencing any dysfunctions in their sexual response cycle. That is, they report desire, arousal and regular orgasms, but with little enjoyment. In these cases, no true sexual dysfunction is present; actions are completed but the encounters are essentially unsatisfying. While a dysfunctional relationship is frequently the cause of such a complaint, at other times the problem represents an example of guilt about pleasure and fear of letting go sexually. For some people the act of letting go sexually involves a focus on the self rather than the partner. This is often labeled as a form of self-centeredness, or even self-love, and can itself inspire guilt because the person may feel that they are using their partner as a means to achieve their own sexual satisfaction. This is a good example of the tension resulting from the tendency of religions to emphasize agape, a selfless love, over eros, a love desiring fulfillment ([17], pp. 126–127). What can happen is that the understanding of the self can be seen in terms of good or bad actions. Once actions are labeled as being bad in the religious sense, it is a simple emotional step to mistake one's actions with one's being and one's feelings with action. Scholastic philosophy told us as much when it echoed Aristotle's metaphysical premise that action follows being.

The lack of sexual satisfaction is just one of many sexual difficulties experienced by couples. Research has shown that sexual difficulties are often more disruptive to relationships than sexual dysfunctions [3]. For example, sexual difficulties arise in relationships in the form of disagreement about sexual habits or practices, amount of foreplay before intercourse, and an inability to relax during sex [30]. These and other difficulties can significantly stress a sexual relationship. I have found that guilt about pleasure has for some people been overcome by embracing denominational teaching about the sacramentality of sexuality in marriage, thus engaging sexuality as the metaphorical marriage between the church as the bride of Christ. However, such an approach for many perpetuates the disembodiment of sexuality from spirituality. Nelson's focus on an incarnational approach to sexuality ([14], [16]) is of great assistance here to those who feel more comfort working within a religious framework.

While I have given attention to the role of religious issues in dealing with problems of sexual desire, sexual satisfaction and before that an approach to homosexuality, there are other areas in which the sex therapist can often come across the effects of moral absolutes when treating patients. Personal moral beliefs are often the issue for the patient when sex therapy deals with intentional sexual acts that are introduced to assist in treatment. Here I am speaking about the use of masturbation, active sexual fantasy, sexually explicit materials (what some would call pornography), sexual surrogates, pre-marital or extra-marital relationships, divorce, abortion, artificial insemination, in-vitro fertilization and gestational carriers as a partial list of possibilities. For many, religious attitudes and belief in moral absolutes forbid specific activities. For others, the ethical principle of double effect can be invoked. This permits doing an evil act as the unintended side effect of a (therapeutically) good act. There are those for whom no moral problem is presented because to them their personal goal is perceived as a good one (the end justifying the means). I have seen this to be particularly true when dealing with infertile couples whose desire to conceive usually eclipses concerns about church teachings [26].

My experience has been that while religious background may at times play a part in the etiology of sexual problems and addressing them is a part of treatment, rarely have I found people unwilling to participate in therapy because the approach is at variance with religious principles. One might say that this is a self-selecting population of usually bright and articulate people who seek treatment. While this is for the most part true, the people in therapy represent a culture that has recently undergone significant changes affecting the individual's world view. Scientific advances such as the birth control pill have changed woman's role as a sexual partner forever. So has the feminist movement, and a more egalitarian view of marriage has made inroads into the patriarchal attitudes present in the history of religion. Modern psychology in the post-Freudian era has placed emphasis on becoming whole persons, a notion that includes sexuality as a good to be enjoyed. Our culture also has come to support the original notion of individual liberty and to appreciate or at least tolerate individual differences. And there are those who, regardless of religious background, think for themselves and are able to experience their sexuality relatively free from the constraints of traditional religious attitudes about sex. I found this last observation to be true for both active church-goers as well as those who declare no active religious affiliation.

RELIGION AS A POSITIVE INFLUENCE

It has not been my intention to view the effects of religious background on sexual adjustment in negative terms. As a clinician I tend to come across problems and the anecdotal clinical impression can be one-sided. However, I have also found religious background to be a positive influence for some people. I have often found that patients for whom affiliation with religion and doctrinal adherence are important are interested in being informed about more modern ways of understanding sexuality within a religious framework. A better understanding of homosexuality is a prime example and comfort has been received by those concerned with their sexual orientation as it relates to their belief system. The holistic approach to sexuality that places emphasis on the person rather than on specific acts makes sense to many people who would otherwise be mired in guilt or feel alienated because of a personal inability to follow or accept traditional teachings.

The power of religion can be used as a valuable resource in permission-giving to those who respect and need the structure and the comfort of feeling within an ethical system. I have admired the flexibility and simplicity with which some people are able to incorporate new religious understanding into their belief system. This is particularly liberating for those couples who are able to expand their sexual script or repertoire during the course of treatment. In such cases early religious and cultural inhibitions resulted in an individual or couple feeling inhibited in expressing a broad range of sexual behavior. What is at work here, I believe, are not specific religious prohibitions about sexual techniques, but a basic mistrust of sexual desire, expression and enjoyment, frequently influenced by sex-negative religious experience.

For example, many couples and individuals may initially register discomfort with common sexual practices such as oral-genital pleasuring; the use of various coital positions; the role of masturbation, both mutual and solitary; the use of sexual fantasy, and sexually explicit talk to enhance arousal and pleasure. They may feel that these and other behaviors are outside of the norms of their religious belief systems. In fact, believers may not be aware that their religious tradition may be quite generous in supporting the expression of sexual love. Often a simple discussion of the religious tradition on sexuality in a permission-giving and supportive fashion brings a sense of relief and confidence to the couple to pursue their sexual relationship joyfully. It is as if a switch were thrown that makes the new behavior "ok." As often seen in other situations, it is the general emotional health of the person that predicts better sexual adjustment, rather than religious observance per se.

In summary, I have found in my work that religious background can be a significant variable in the sexual adjustment of many of the persons I have treated over the years. For the most part, the influence of religion was negative and at times represented an obstacle to be overcome either through a better understanding of official teachings, a demythologizing of erroneous beliefs and influences, or even outright rejection of religious teachings on sexuality. For some people, religion was so important that sexual issues had to be dealt with within a religious framework to guarantee successful treatment. Others were able to leave religious teachings "outside the door" of the therapist's office and felt comfortable in dealing with sexuality at a personal, experiential and contemporary-scientific level.

A final category of patients expressed a feeling that religion had no bearing on their past or current sexual attitudes or functioning, except that they recognized that they were raised within the influences of our society and were aware of societal norms. My concluding feeling is that regardless of the therapist's own religious background or current relationship with organized religion, he or she would be prudent to pay careful attention to the influence that religious factors have on the patient's attitude about sexuality. It is too easy for the therapist to hide behind the pretense of scientific objectivity and in doing so miss an important if not in some cases essential component of the psychological character of the one who seeks assistance.

Pennsylvania Hospital
Philadelphia, Pennsylvania, U.S.A.

BIBLIOGRAPHY

1. English, O.S. and Pearson, G.H.: 1963, *Emotional Problems of Living*, W.W. Norton, New York.
2. Fortunato, J.E.: 1987, *AIDS, the Spiritual Dilemma*, Harper & Row, San Francisco.
3. Frank, E. et al.: 1978, 'Frequency of Sexual Dysfunction in "Normal" Couples', *New England Journal of Medicine*, 299, 111–115.
4. Gramick, J. and Furey, P. (eds.): 1988, *The Vatican and Homosexuality*, Crossroad, New York.
5. Green, R.: 1987, 'The Irrelevance of Theology for Sexual Ethics', in E.E. Shelp (ed.), *Sexuality and Medicine*, Vol. II, D. Reidel, Dordrecht, Holland, pp. 249–270.

6. Kaplan, H.S.: 1979, *Disorders of Sexual Desire,* Brunner/Mazel, New York.
7. Kaplan, H.S.: 1987, *Sexual Aversion, Sexual Phobias, and Panic Disorder,* Brunner/Mazel, New York.
8. Kosnik, A. et al.: 1977, *Human Sexuality: New Directions in American Catholic Thought,* Paulist Press, New York.
9. Leiblum, S.R. and Rosen, R.C. (eds.): 1988, *Sexual Desire Disorders,* Guilford Press, New York.
10. Masters, W. and Johnson, V.: 1970, *Human Sexual Inadequacy,* Little, Brown, Boston.
11. McNeil, J.: 1985, *The Church and the Homosexual,* New Year Publications, New York.
12. McNeil, J.: 1988, *Taking a Chance on God,* Beacon Press, Boston.
13. Money, J.: 1990, *Address to the American Association of Sex Educators, Counselors and Therapists,* Washington, D.C., (March).
14. Nelson, J.B.: 1978, *Embodiment: An Approach to Sexuality and Christian Theology,* Augsburg Publishing House, Minneapolis.
15. Nelson, J.B.: 1982, 'Religious and Moral Issues in Working with Homosexual Clients', in J. Gonsiorek (ed.), *Homosexuality and Psycho-therapy,* Haworth Press, New York, pp. 163–175.
16. Nelson, J.B.: 1983, 'Sexuality Issues in American Religious Groups: An Update', *Marriage and Family Review* 6:3/4, 35–46.
17. Nelson, J.B.: 1983, 'Religious Dimensions of Sexual Health', in G. Albee, S. Gordon and H. Leitenberg (eds.), *Promoting Sexual Responsibility and Preventing Sexual Problems,* University Press of New England, Hanover & London, pp. 121–132.
18. Nelson, J.B.: 1983, *Between Two Gardens, Reflections on Sexuality and Religious Experience,* Pilgrim Press, New York.
19. Nelson, J.B.: 1987, 'Reuniting Sexuality and Spirituality', *Christian Century* 2/25, 187–190.
20. Nelson, J.B.: 1988, *The Intimate Connection: Male Sexuality, Masculine Spirituality,* Westminster Press, Philadelphia.
21. Pagels, E.: 1988, *Adam, Eve and the Serpent,* Random House, New York.
22. Scroggs, R.: 1983, *The New Testament and Homosexuality,* Fortress Press, Philadelphia.
23. Sedgewick, T.: 1989, *Christian Ethics and Human Sexuality: Mapping the Conversation,* (unpublished manuscript).
24. Shelp, E.E. and Sunderland, R.H.: 1987, *AIDS and the Church,* Westminster Press, Philadelphia.
25. Slowinski, J.W.: 1974, *Impotence in the Male: A Review of Treatment Forms,* Unpublished Masters Thesis, Hahnemann Medical College, Philadelphia.
26. Slowinski, J.W.: 1987, 'The Infertile Couple: Issues in Treatment', *Presentation: Society for the Scientific Study of Sex,* Atlanta 11/6/87.
27. Slowinski, J.W.: 1988, 'Persons with AIDS and Religion: A Clinician's Viewpoint', *Society for Sex Therapy and Research Newsletter* 6:4, 4.
28. Slowinski, J.W.: 1988, 'AIDS, Gays and Religion: Reflections on Dealing with the Religious Beliefs of HIV Positive and AIDS Patients', *Presentation: Society for the Scientific Study of Sex,* San Francisco, 11/5/88.
29. Slowinski, J.W.: 1989, 'Psychological Needs of HIV Positive and AIDS Patients', *Medical Aspects of Human Sexuality* 23 (9), 52–54.
30. Slowinski, J.W.: 1989, 'Sexual Dysfunctions: Common Problems Often Overlooked', *Pennsylvania Hospital OB/GYN Medical Newsletter* 2 (1), 3.
31. Slowinski, J.W.: 1989, 'Sexuality and Spirituality: Historical Roots in the Western Christian Tradition', *Presentation, Society for the Scientific Study of Sex,* Toronto, Canada, 11/10/89.
32. Stackhouse, B.: 1990: 'The Impact of Religion on Sexuality Education', *SIECUS Report* 18 (2), 21–24.
33. Yates, W.: 1988, 'The Church and Its Holistic Paradigm of Sexuality', *SIECUS Report* 16 (4–5), 1–5.

WILLIAM S. SIMPSON AND JOANNE A. RAMBERG

THE INFLUENCE OF RELIGION ON SEXUALITY: IMPLICATIONS FOR SEX THERAPY

In reviewing our work with couples in sex therapy during the past few years, we have noted a striking link between a highly religious upbringing and sexual dysfunctioning, with concomitant resistance to the proven techniques of sex therapy. This resistance encountered in sex therapy is apparently magnified by the watchful superego that devout individuals often develop at a young age. Thus sex therapists should initially address the interface of religious beliefs and attitudes about sexuality, and should incorporate an ongoing discussion of this dynamic into the therapy process. A computerized search of the sex therapy literature revealed a lack of focus on this critical influence on sexual development, with only generalized statements about how religious beliefs affect sexual functioning and, subsequently, sex therapy.

Our experience in treating persons whose religious background is traditionally conservative shows that they tend to lack basic knowledge of sexuality in general. They are also fairly inexperienced sexually, even in regard to self-exploration. Although these limitations on their sexuality seem to be self-imposed, they stem in part from a lifelong lack of openness in the family about sexual matters, as well as from a dominant religious overtone of prohibition and censure.

As do most sex therapy patients, these individuals require basic education about human sexuality. Thus, our treatment methods involve traditional behavioral modification techniques of desensitization and sensate focus exercises [7]. To further inform the patients, we provide educational books [6], [7], [11], [15], and show explicit videotapes [5], [10], [12], [13], [14]. In addition to providing reassurance that enjoyable sexual behavior is acceptable, sex therapists can use biofeedback exercises and relaxation training to help such patients alleviate their acute performance anxiety.

Professionals in the field of sex therapy must develop a set of personal values that is both respectful and non-judgmental. The ethical code of the American Association of Sex Educators, Counselors and Therapists (AASECT) encourages therapists to respect the client's right to hold values differing from those of the practitioner, and also assigns therapists the responsibility for assessing and working within the client's values [2]. One value-laden influence on an individual's sexual behavior that is often overlooked or not emphasized in sex therapy is religion. Yet Bloomfield and Marteau [1] have stressed that religious beliefs significantly influence attitudes about sexuality, and that those who want to conform to a rigid religious code of conduct may struggle to suppress otherwise healthy sexual impulses and desires that do not conform.

Since religion is a very personal aspect of human lives, it is bound to play a

Ronald M. Green (ed.), Religion and Sexual Health, 155–165.

role in determining sexual values and behavior. In fact, Calderone [3] has called attention to the influence that religious institutions have on sex education in all primary and secondary school systems. And developing a scholastic sex education curriculum requires some knowledge of the cultural values of the student population. Gordon and Snyder [4] distinguished between sex and sexuality, defining *sex* as a biologically based need and *sexuality* as one's self-understanding as a male or female and as a means of communicating with other humans.

Similarly, Masters and Johnson ([8], [9]) have noted that a rigid religious childhood background is often associated with many sexual dysfunctions. Although they linked several specific dysfunctions with such a background, they qualified their thesis by stressing that the fault lay in the severely antisexual attitude forced on the child rather than with the religious beliefs per se. To aid couples whose sexual difficulties involve conflict between their religious teachings and values and their human desires and impulses, we often prescribe the reading of the book, *The Gift of Sex* [11].

Religious beliefs may play a strong role in the presenting problem itself, or may become a significant resistance that must be dealt with in the treatment process. More commonly though, religious beliefs seldom play a primary role. Sexually dysfunctional individuals or couples whose problems can be linked to their religious upbringing may not seek sex therapy in the first place, however, because of the overwhelming resistance to taking just this initial step. Such resistance is strong even in persons without strong religious constraints.

Thus the following clinical case examples may shed some light on resistance in general and, specifically, on religious-based inhibitions to seeking and participating in sex therapy. Unfortunately, sex therapists do not practice in an ideal clinical world. People who make an initial appointment for sex therapy often cancel it for no reason. Others will participate in an initial consultation and will establish a therapeutic plan that may not be followed through, or they may begin a therapy process only to drop out of it for no apparent reason. As a result, sex therapists are frequently left feeling helpless because more could have been done if the opportunity had arisen.

CLINICAL VIGNETTES

Case No. 1: Mr. and Mrs. Adams

A carpenter from a rural area and his homemaker wife were referred by their family doctor for sex therapy after the husband had suddenly become unable to get and maintain erections, resulting in depression and a loss of interest in sexual activity. Both Mr. and Mrs. Adams were in their early forties and reported having had a satisfactory sexual life throughout their 20-year marriage. The precipitating event for the husband's impotence appeared to be his wife's revelation that she could only achieve orgasm by fantasizing during intercourse, or by masturbating herself in private. The fantasy that aroused her was one of being spanked lightly on the bare buttocks.

Both the husband and the wife had been reared in a conservative fundamentalist religious family, and both continued to live as devout Christians. Mr. Adams noted that his faith had been strengthened when he had recovered from a severe illness that struck him during adolescence. Because his life had been spared, he rededicated himself to leading an exemplary existence, part of which involved his total and irrevocable abnegation of masturbation.

Unlike her husband, Mrs. Adams had maintained an enjoyment in masturbation, which she recalled learning as a preschooler. About the same time, she had been sexually molested by a teenaged neighbor, who had fondled her and had stroked her vulva to arouse her. Her spanking fantasy apparently developed after her father found her masturbating and punished her by striking her bare buttocks several times with the palm of his hand. Mrs. Adams was subsequently able to use the spanking fantasy to arouse herself to a high level of sexual excitation. Although she continued to use this fantasy after she married, she had been reluctant to tell her husband about it.

Since Mr. Adams did not approve of masturbation himself, he somehow felt that it was sinful for his wife to masturbate and to use the fantasy of being spanked. He also thought that he was less of a man because he was unable wholly to satisfy his wife's sexual needs. Her revelation created such intrapsychic conflicts for him that he quickly developed a major depression. Mr. Adams' distress and depression led him to consult a therapist in private practice who was also a member of his church. When his depression failed to improve, Mr. Adams saw his family doctor, who referred the couple to us for sex therapy. Their initial treatment plan included a prescribed antidepressant (fluoxetine hydrochloride, 20 mg daily) for Mr. Adams.

Both Mr. and Mrs. Adams indicated that they had been virgins on marriage, and had not had any other sexual partners since. As a result of their conservative background and restricted experience, they were very ignorant about human sexuality in general. We encouraged them to view both masturbation and fantasizing as part of normal, permissible sexual behavior. They seemed quite relieved by this knowledge and by our explanations that it was perfectly all right to use whatever means suited a person to increase sexual pleasure.

As they became more knowledgeable about – and more accepting of – their sexuality, they were able to perform sexual intercourse in the female superior position for the first time in their marriage. Mrs. Adams noted that she had become quite aroused when her husband had touched her buttocks during intercourse in the new position. They also began to openly explore each other's body and to communicate to each other about what was or was not pleasurable. Over time, Mrs. Adams was able to incorporate her masturbation into the act of intercourse itself, without either her or her husband feeling that doing so was sinful. They even were able to accept our recommendation to use a vibrator together.

Meanwhile, Mr. Adams responded well to the traditional behavioral techniques for overcoming erectile problems. Within three months from the time of their first session, they reported that they had begun to have the best sex of their marriage. For example, they were able to have mutual simultaneous orgasm after only seven sessions of sex therapy.

Comment: This case illustrates how individuals who hold extraordinarily strong religious beliefs can develop a severe psychological reaction on discovery of a so-called sinful component to the sexual arousal process. Yet it also reveals how quickly a resolution of intrapsychic conflict can occur in an accepting therapeutic atmosphere. Such couples may find that sex therapy not only addresses the initial problem, but that it also enhances their overall sex life.

Case No. 2: Mr. Brown

This 40-year-old, divorced man was referred for sex therapy because of erectile difficulty for which his physician could find no physical cause. He had only recently been divorced after 18 years of marriage, and he had a teenage daughter. Mr. Brown was a good-looking, well-educated, intelligent man who held a high leadership position within a religious organization that emphasized family values, work ethic, and fidelity, and in which divorce was very much frowned upon. He was a dedicated, honest man who adhered closely to the strictures of his church.

Initially, his marriage had been very satisfying to both him and his wife, but as his responsibilities in his religious vocation increased, he began to spend enormous amounts of time away from home. More and more of the responsibility for rearing their daughter fell to Mrs. Brown, who began to develop a life of her own and to grow more independent. As their daughter got older, Mrs. Brown resumed the professional work she had been engaged in before her marriage, and rose rapidly within her field. As Mr. Brown continued to travel and to be devoted to his work, his wife suddenly announced that she had been unhappy with their marriage for several years, that she was filing for divorce, and that nothing he could do would stop her. Although Mr. Brown reluctantly assented to the divorce, he never ceased loving his wife. His conscientiousness demanded of him that he provide even more child support than required by law, and he also provided his ex-wife with an extraordinarily large portion of the material acquisitions that their shrewd investments had secured.

Although Mr. Brown indicated that there had been no problems in his sexual functioning with his former wife, he was currently unable to get and maintain erections during times of intimacy with his present girlfriend. She was a woman whom he greatly respected and admired and to whom he felt sexually attracted. He was, however, unable to function with her as a sexual partner because he viewed intercourse outside marriage as a sin.

Mr. Brown appeared to be a rather sad, mildly depressed man, very much at odds with his own harsh conscience. Although the initial diagnostic session helped clarify for him the nature of his conflict, he proved resistant to treatment. He made an appointment for another interview, which was to be the first of several sessions, but he later called to cancel it and never rescheduled.

Comment: This case illustrates how unresolved grief over the loss of a former spouse can combine with a harsh superego to produce seemingly unresolvable internal conflict. Although such patients might be helped, they often simply do not give sex therapy a chance, instead terminating the process prematurely.

Case No. 3: Mr. Charif

This Indian man in his mid-thirties, reared in Britain, had moved to the United States when he became a graduate student. Under the stress of a new culture, academic pressure, and family expectations, he became ill and required hospitalization. He was referred for sex therapy because he complained of having recently developed an inability to get and maintain erections. Quite apart from this physical problem, he was found to be suffering from a major psychiatric disorder (i.e., atypical psychosis, with paranoid features). He also had some mild neurological sequelae as a result of a fall he had taken on his bicycle a couple of years earlier.

Mr. Charif's earliest sexual activity had been confined to fondling his penis until he had an erection, but without orgasm. (Masturbation, per se, was forbidden by his religion.) Later on, Mr. Charif had been able to function sexually with various American women as his sexual partners. His sexual dysfunction first surfaced when his father announced to him that a marriage had been arranged to a young woman back in India whom he had met only once. His father would not even allow him to consider the possibility of marriage to a woman outside their faith or country of origin.

Mr. Charif found himself trying to avoid the arranged marriage by any conceivable means. He pleaded with the hospital staff to tell his family that he was not ready for marriage to anyone. At the time of the consultation, however, he was imminently ready to be discharged from the hospital to return to his family in another state. In spite of the staff's communication to the patient's father that a marriage should not be contracted, the patient indicated, dejectedly and hopelessly, at discharge that he thought he would have to go home and get married.

Comment: This case illustrates the limited value of a brief sex therapy consultation with a psychiatrically disturbed male patient faced with his own and his family's inflexible religious beliefs. Although Mr. Charif felt helpless in the face of his father's demands, the therapist felt equally helpless.

Case No. 4: Mr. and Mrs. Din

This 30-year-old man and his 22-year-old wife were referred by their family doctor for treatment of what appeared to be the husband's premature ejaculation. In the initial interview, however, we found that although Mr. Din did ejaculate before penetration, it was because he had never been able to penetrate his wife's vagina. Her severe vaginismus had prevented the consummation of their one-year marriage. Both had been brought up in an Asian country where morality is very strict, particularly for women. They had both immigrated to the United States to attend the university where they met. As was customary in their culture, the husband had had many previous sexual experiences (without any sexual dysfunctioning), but the wife was a virgin and had very limited information about sexuality, as mandated by their religion. In addition, she had been brought up to believe that intercourse was painful for women, so the pain she had felt on her wedding night when her husband attempted penetration reinforced her fear.

The wife was referred to a female gynecologist (she refused to be examined by any male physician), who determined that she had a stenotic hymen that would not yield to digital pressure. A hymenotomy was performed and the vagina was easily dilated. After Mrs. Din's hymenotomy had healed, she returned to sex therapy for treatment of her vaginismus and was lent a set of vaginal dilators, which she was encouraged to use to gradually increase her ability to dilate. After this session, however, the Dins did not return for further treatment. They scheduled several appointments, only to cancel each one. Finally, after a hiatus of about three months, Mrs. Din called to say that she and her husband had been able to have sexual intercourse successfully, and that she would return the dilators. However, she went on to say that she was now totally unable to have intercourse because she was so depressed. Her mother had recently died and she was undergoing a religiously mandated period of mourning and bereavement. She was told that she should not discontinue treatment, but instead should come in right away for treatment of her severe depression because we were qualified to treat this illness as well. She said she would have to talk it over with her husband. We never heard from her again.

Comment: This case illustrates the confusion that can surround the referral of a patient for a sexual disorder when the referring physician himself does not get an adequate history from the patient of his sexual complaint because of the physician's own reticence in discussing sexual matters. The sex therapist's diagnostic interviews themselves were therapeutic, and resulted in further examination and treatment that led to a successful outcome. Also, this case illustrates how a patient can develop a sexual dysfunction due to depression, but might refuse treatment for the depression when the presence of inadequate sexual desire with bereavement is congruent with religious beliefs.

Case No. 5: Mr. and Mrs. Evans

This married couple, both 27 and both successful academicians, were referred because of the husband's severe erectile difficulties. Mr. Evans had been treated by a urologist with intracorporal papaverine-phentolamine injections, which had been successful in producing erections; however, when he and his wife then had intercourse, he had been unable to ejaculate. After a few months of marital therapy, they had expanded their sexual knowledge, but their sex relations remained unimproved. Mrs. Evans noted that her husband didn't seem interested in sex, although they had been married only a year. During their attempts at intercourse, for example, he manifested none of the expected signs of sexual arousal.

Mr. Evans had been reared by extraordinarily conservative Catholic parents, who provided him with no sex education, and discouraged him from obtaining information about sex from peers or books. His parents were in their forties when he was born, and they had always been very reserved about their own sexuality; as a child, the patient incorporated this restrictiveness. As a preadolescent, he took seriously the parish priest's admonition that masturbation was a sin, so did not learn to masturbate until he was married and only then as part of his education during treatment. Until he learned to masturbate, Mr. Evans had never had an orgasm, and he could not recall ever having had nocturnal emissions.

During his high school years, Mr. Evans did not date because he was so shy, serious, and studious. In college he had tried to date but without much success. He had met his wife on a blind date and had courted her for almost three years. Much of this time was spent in non-sexual companionship. It was months before he even kissed her, and only then at her initiative. After they became engaged, he continued to resist any sexual activity prior to their wedding because of his religious beliefs.

Although Mrs. Evans had been reared in a conservative Protestant family, she was not as sexually restricted. Like her husband, she had never masturbated, but she had been able to be orgasmic during intercourse with him after the papaverine-phentolamine injections. Since Mr. Evans had so actively avoided learning anything about sex throughout his life, he was quite ignorant about sex in general. He was open, however, to reading about human sexuality, particularly in men, and to practicing sensate focus exercises with his wife. They both expressed surprise at how much there was to learn. After carrying out the sensate focus exercises for a period of three months, Mr. Evans was able to get and maintain an erection whenever his wife stroked his penis.

Despite their success in producing and maintaining erections using the classical behavioral techniques of Masters and Johnson ([8], [9]), Mr. Evans continued to be unable to ejaculate intravaginally with intercourse alone. He also expressed little sexual desire, even indicating that he was not stimulated by intercourse but instead felt only wetness and warmth when his penis was inside his wife. He could masturbate to ejaculation, but it took a long time and required intense stimulation. His breathing rate increased only when he ejaculated.

The treatment was interrupted for a month when Mrs. Evans became acutely and seriously ill. When faced with the threat of losing his wife, Mr. Evans became much more attentive to her on her recovery. They achieved their greatest level of sexual success about a week later. His newfound sexual interest soon plummeted, however, and he appeared again to participate only reluctantly in the sex therapy assignments. His wife noted that he seemed to regard them as "chores" to be done. In fact, Mr. Evans appeared pleased at having little or no sexual desire. Yet when we pointed out this reaction formation to him, he realized that his wife had been quite patient with his sexual dysfunction, and that if he were to lose her, he might have enormous difficulty recruiting another sexual partner. Using biofeedback and relaxation training, Mr. Evans began practicing a hierarchical exercise to overcome his anxiety about sexual performance. But, again, there was no perceptible therapeutic effect.

Mr. Evans also seemed to view his physical and interpersonal clumsiness with women as evidence that he was "pure," in comparison to the "smooth" young men he had noticed as he grew up. His secret narcissistic pride in being awkward served to strengthen his repression and to bolster his denial of his own sexuality. We encouraged him to own his own sexuality, which in the past had been managed by his parents, his priest, and others.

The patient's erectile difficulties and his associated anxiety continued to be so severe that we prescribed a vacuum pump device to help him to get erections and to have intercourse, thus bypassing his anxiety. Almost a month passed before he would even use it. Mrs. Evens enjoyed the experience thoroughly and was multiply orgasmic, so she became enthusiastic about the use of the device. Mr. Evans, however, complained that he felt like a "piece of meat," gained no pleasure from the experience, and was still unable to ejaculate intravaginally with intercourse alone.

About a year into the sex therapy process, Mr. Evans was able to allow himself to have fantasies about an actress, and his wife encouraged him to use the fantasies. He was also able to achieve orgasm through masturbation within a few minutes. To achieve intravaginal ejaculation, however, either Mr. Evans or his wife had to masturbate him past the point of ejaculatory inevitability, then quickly insert his penis into her vagina. With practice, Mrs. Evans was able to gauge exactly when to insert the penis so that ejaculation could take place intravaginally about 95 percent of the time. Meanwhile, Mr. Evans reported that his genital sensitivity seemed to be increasing. By session 10, about halfway through their treatment, they were having intercourse twice a week, and he was showing signs of increased sexual arousal. Because of their desire for children, they were planning to increase the frequency of their intercourse to three times a week during Mrs. Evans ovulatory periods. However, they were never able to achieve intravaginal ejaculation with intercourse alone.

Comment: This case illustrates how a repressive, restrictive sexual environment and religious upbringing can cause a primary sexual dysfunction that dominates adult sexuality. Because repression, denial, and reaction formation serve to perpetuate anachronistic beliefs and behavior, they produce redoubtable resistance to sex therapy.

Case No. 6: Mr. and Mrs. Frank

This 37-year-old man and his 34-year-old wife, both of them in their second marriage, were referred to us for treatment of the husband's inability to get and maintain erections. Mr. Frank had had no sexual dysfunction before or during his first marriage. During his courtship of his present wife, he had had some erectile difficulty, but nothing that they thought they should be concerned about. The episodes of erectile difficulty became more frequent after their marriage, however. Increasingly, his erection would disappear after intromission, with concomitant inability to ejaculate. This development caused his wife to feel rejected and to worry that she was sexually unattractive to him.

As Mr. Frank's sexual difficulties increased, he became less interested in foreplay, which became abbreviated and therefore less mutually satisfying. Mrs. Frank eventually felt so rejected that she contemplated filing for divorce, but instead convinced her husband to seek help. She made the initial contact with us, after which several weeks passed before he would set up an appointment. By the time the two of them came to their first session, they had separated. Although Mrs. Frank still seemed interested in developing a good sex life with her husband, her own sexual insecurity had skyrocketed when he complained to her that perhaps he would be able to perform better with another woman.

In the initial interview, it became obvious that both Mr. and Mrs. Frank were quite depressed. Both had vegetative signs of depression, so fluoxetine hydrochloride (20 mg daily) was prescribed for each. Within three weeks, they both responded well to the antidepressant medication. They also were quite successful in performing the nongenital, nonbreast touching exercises of the sensate focus treatment we recommended. But as genital and breast stimulation was added to the sensate focus regimen, Mrs. Frank began to express mixed emotions about the therapy. She became even more upset at the prospect of intercourse after her husband was able to obtain and maintain erections consistently.

At this point, Mrs. Frank revealed information about her mother's second marriage. For 15 years, her mother had been married to a man whose religion espoused polygamy. This stepfather had also made sexual advances to her when she was an adolescent. In particular, when her mother was away at work he would talk to her about the primum noctus medieval doctrine of fathers initiating their daughters into sexual relations. He had stopped this activity, however, when she told her mother about his actions.

Nevertheless, as a young girl Mrs. Frank had become accustomed to the religious belief that wives had to submit to "God's will" and marry a second man if so ordered. This revelation led us to interpret her concern about her husband's "threat" to find another woman with whom he could perform better sexually. Mrs. Frank had immediately concluded, albeit unconsciously, that perhaps she would become a second – and less important – wife.

In addition, her first husband had been unfaithful to her, so she had mixed feelings about the sex therapy itself. On the one hand, she very much wanted her husband to conquer his sexual difficulties, but on the other hand, if he got better, he might find another partner altogether. Fueling this worry was the fact that her stepfather had been a philanderer and had fathered a child by another woman during his marriage to her mother.

When this conflict was brought into consciousness, the couple was able to talk rationally about it, and Mr. Frank assured his wife that he had no desire for sex with anyone but her. They continued their prescribed behavioral exercises, and soon became totally sexually functional together. We gave them the option of stopping the therapy at this point, with a subsequent appointment in a month to see how they were doing. Shortly before the scheduled follow-up appointment, Mrs. Frank called to say that the two of them were continuing to function together sexually to their satisfaction, so the appointment was cancelled.

Comment: This case illustrates that a couple may seek treatment, and even participate in the early stages of treatment, without revealing any underlying religious factors. Only when obvious resistance (in the form of sabotaging treatment) appears do religious factors sometimes surface, and these must be addressed in order for therapy to produce a successful outcome.

CONCLUSION

Religious beliefs are only one of the many factors that determine whether sexually dysfunctional persons will seek out a sex therapist in the first place. Individuals in need of sex therapy may hesitate to seek it or use it fully because of inhibitions inculcated by their religious upbringing. Religious beliefs also influence one's ability to make good use of the therapy. They may be cited as a reason to resist therapeutic efforts or may influence patients to drop out of therapy for no reason despite the process being incomplete.

Our purpose herein has been to make sex therapists more aware of this dynamic and to help address the paucity of clinical case material in the literature. As others in the field review their own experiences in treating individuals with strong religious beliefs, they may become better able to identify developmental patterns and their consequences and to deal with them earlier in the treatment.

When individuals with a deeply ingrained resistance to discussing sexual matters nevertheless attempt to seek some alleviation of a sexual dysfunction, it indicates that they are strongly motivated to change. So, too, does the fact that they often enter sex therapy on the advice of a more general medical practitioner (e.g., a family physician) who has been unable to resolve the sexual dysfunction. An increased sensitivity to the patient's religious beliefs on the part of the sex therapist can enhance the likelihood of bringing the treatment process to a successful resolution.

Center for Sexual Health, The Menninger Clinic
Topeka, Kansas, U.S.A.

and

Washburn University
Topeka, Kansas, U.S.A.

BIBLIOGRAPHY

1. Bloomfield, I. and Marteau, L.: 1976, 'Psychosexual Problems in a Religious Setting', in S. Crown (ed.), *Psychosexual Problems,* Grune and Stratton, New York, pp. 265–296.
2. Brown, R.A. and Field, J.R. (eds.): 1988, *Treatment of Sexual Problems in Individuals and Couples Therapy,* P.M.A. Publishing Corp., New York.
3. Calderone, M.S.: 1976, 'Education for Sexuality', in B.J. Sadock, H.I. Kaplan, and A.M. Freedman (eds.), *The Sexual Experience,* Williams and Wilkins Co., Baltimore, pp. 518–526.
4. Gordon, S. and Snyder, C.W.: 1989, *A Guidebook for Better Sexual Health,* 2nd ed., Allyn and Bacon, Boston.
5. Heiman, J. and LoPiccolo, J.: 1976, *Becoming Orgasmic* (Parts I, II, III), [videotape]: Multi-Focus, Inc., New York.
6. Heiman, J. and LoPiccolo, J.: 1988, *Becoming Orgasmic: A Sexual Growth Program for Women,* Prentice-Hall, Inc., Englewood Cliffs, New Jersey.
7. Kaplan, H.S.: 1979, *Making Sense of Sex,* Simon and Schuster, New York.
8. Masters, W.H. and Johnson, V.E.: 1970, *Human Sexual Inadequacy,* Little, Brown and Co., Boston.
9. Masters, W.H., Johnson, V.E. and Kolodny, R.C.: 1986, *Masters and Johnson on Sex and Human Loving,* Little, Brown and Co., Boston.
10. *The Miracle of Life* [videotape]: 1986, NOVA, Sveriges TV/WGBH, Boston.
11. Penner, C. and Penner, J.: 1981, *The Gift of Sex,* Word Books, Waco, Texas.
12. *Sensate Focus Exercises* (Parts 1 through 4) [videotape]: 1975, Ecoa Productions, Avenue Video, New Jersey.
13. Wagner, G.: 1974, *Physiological Reactions of Sexually Stimulated Female in the Laboratory* [videotape], Multi-Focus, Inc., New York.
14. Wagner, G.: 1974, *Physiological Reactions of Sexually Stimulated Male in the Laboratory* [videotape], Multi-Focus, Inc., New York.
15. Zilbergeld, B.: 1978, *Male Sexuality,* Bantam Books, New York.

HAROLD I. LIEF

SEXUAL TRANSGRESSIONS OF CLERGY

INTRODUCTION

The sexual exploitation of patients or clients by health professionals has attracted much attention during the last decade, alarming professionals and public alike. The violation of generally agreed upon boundaries between counselor (physician, therapist) and counselee is paralleled by similar violations of the accepted and acceptable roles of the clergy. The public seems to be even more offended by violations of sacramental vows than by breaches of conduct set forth in the Oath of Hippocrates.

Why should this be so? Why should the clergy be held to a higher standard of behavior than, let us say, a physician? In part, it depends on the nature of the transgression. If both doctors and clergymen (in our discussion we will be talking only about men; the issue of lesbian clergy, for example, important as it is, will be left out) were engaging in pedophilia, their behavior probably would offend equally. On the other hand, adulterous and gay behavior have quite different valences for clerics and health professionals. Why? With regard to the Roman Catholic priest the answer is obvious. The requirement for celibacy makes any sexual act, even the most loving and consensual, a violation of a sacred oath, therefore, automatically a transgression.

That Roman Catholic priests are held to a higher standard because of the demand for celibacy accounts for the public outrage when a Father Bruce Ritter of Covenant House was found to have had sexual relations with youths under his care.[1]

What about clergy for whom celibacy is not a requirement? If the priest or minister is to guide his flock in the path of virtue he must be a role model. The congregation elevates the cleric, puts him on a pedestal, gives him a cloak of purity and goodness, probably derived from images of Christ and Moses, the great teachers and prophets who were able to transcend their own temptations and self interest. The expectations of the congregation are heightened by virtue of the public role of the minister, for he is symbolic of the Bible in ways that lawyers, for example, do not symbolize the Law.

An additional dimension of the expectations of the public relates to the central role of the family in the Judeo-Christian theology and way of life. The Bible reiterates the sanctity of the home and family; this is so central that it may account in part for the requirement of celibacy in Roman Catholic tradition. The conflict between family and priesthood was noted by Christ himself and led St. Paul and others to mandate celibacy in order to avoid the conflict. (The film

Ronald M. Green (ed.), Religion and Sexual Health, 167–186.

"The Last Temptation of Christ" was at least as much about the pleasures of having a family as engaging in sex.)

These expectations, unrealistic as they are, find reciprocal support in church authority and administration. Either explicitly or tacitly, church authorities expect that their churchmen will uphold the values of the family and aspire to an approximation of purity and goodness.

That these expectations by church authority and the public create enormous strain and conflict on priests, ministers, and rabbis has formed the basis for fascinating fiction. The wayward clergyman has been a favorite of many modern writers, a few examples of whom are Sinclair Lewis in *Elmer Gantry*, Somerset Maugham in *Rain*, and John Updike in *A Month of Sundays*. In this connection it seems as if truth is really stranger then fiction, for, in the last several years, we have witnessed a number of "fallen preachers," notably, the two Jimmys, Swaggart and Bakker.

These are among the reasons, other then interpretation of the scriptures, that adultery and homosexuality, both seen as threats to family values and stability, are perceived as transgressions, and create potential problems for the clergyman. The church as a social institution now is going through a period of upheaval as it tries to come to terms with current changes in social mores, that include, as well, the issue of sexism. Much of aberrant sexuality involves abusive power, the male exploitation of the female found, in its more extreme form, in sexual addiction or compulsion, an occasional aberration found in the clergy as it is in the general population.

A more frequent abuse of power occurs when the minister (or physician) uses his helping/pastoral role to seduce his parishioners or patients.

It should be clear from this that the standards of the church (these vary, of course, often considerably, from church to church) are often different from the general norms of society so that an adulterous or gay clergyman has special and unique problems that the lay person with similar behavior would often escape.

ROLE AS CONSULTANT

During the decade of the '80s, I served as a psychiatric and sexologist consultant for the Episcopal Church. Priests with sexual problems were referred for diagnosis and recommendations; the latter would consist of treatment recommendations and often suggestions about the priest's future role in the church. The recommendations were usually sent to Bishop David Richards, the source of referral for most of the priests. Bishop Richards headed the Office of Pastoral Development of the Episcopal House of Bishops. His role was to counsel "problem" priests usually referred to his office by the members of the House of Bishops. Problem priests were those whose behavior had created conflict with parishioners, vestry, other priests, or church administration. Several of the priests referred were, themselves, bishops. In most cases, Bishop Richards was able to handle the situation without outside consultation. When he thought the situation required a sex therapist, he turned to me. On a few occasions as a psychiatrist, I was asked to interview priests with problems that had nothing to do with sex; most, however, were sex related and it is these that I wish to review here. (All but one of the eighteen priests I saw came through Bishop Richard's office or through his recommendation.)

During the time that I was engaged in this consultant work for the Episcopal Church, Dr. Julian Slowinski, clinical psychologist, moved into an office next to mine, and I asked him to join me in the consultation process.

PATIENT POPULATION

Of the eighteen priests whom I saw in consultation, fifteen created problems for the church because of their sexual behavior: seven who were (or thought to be) gay; five accused of adultery; three with paraphiliac (deviant) behavior. Of these three, one was a pedophile, another a transvestite, and the third had a sexual addiction. The group of fifteen does not represent a "sample." We have no data to substantiate a belief that this is a representative group of priests with sexual "problem behavior," nor can we extrapolate from this to begin to judge the prevalence of these types of behavior in Episcopal priests or in the clergy in general.

Yet others have made rather startling statements about the prevalence of homosexualities in Roman Catholic priests. (Although homosexuality is not considered a disorder by medical authority, it is not widely accepted as normative by most churches, and certainly not by the Catholic church.) Sipe, basing his report on the clergy he has studied, states:

> Generally, 18% to 22% of clergy (estimates established from all sources) are either involved in homosexual relationships, have a conflict about periodic sexual activity, feel compelled toward homosexual involvements, identify themselves as homosexual, or at least have serious questions about their sexual orientation or differentiation. Not all of these men act out any sexual behavior with others .... Between 1978 and 1985 the reporting of homosexual behaviors increased significantly and the reliable estimates almost doubled to between 38% and 42% ([5], pp. 53ff.

Sipe adds that the reports from several dioceses claim that homosexuality "approached 50%" or even "75% in two dioceses" ([5], p. 56). Sipe assigns the increase to both the feminist movement and the gay liberation movement that together have created greater openness in the expression of sexual affection. "The open expression of the homosexualities in the clergy community, the greater tolerance of individual behaviors, the freedom of movement that makes various lifestyles possible and the increasing need to recruit more priests which has altered admission standards to seminaries and religious houses have all increased the appeal of the priesthood to some who openly identify themselves as gay" ([5], p. 57). He concludes by saying that the *majority* of clergy "could be involved in the homosexualities over the next two decades" ([5], pp. 57). (It needs to be noted that Father Andrew M. Greeley and others have serious reservations about the nature of Sipe's data collection and conclusions.

So far as I am aware, there are no comparable data in the Episcopal Church or in other Protestant denominations. What to do about gay priests is an issue that the Episcopal Church is struggling to resolve. In another section, I will discuss the experiences of gay priests and their difficulties in fitting in, as well as the difficulties the church has in fitting them in. I will also discuss, in other sections, the issues of adultery, of sexism, and of sexual deviation (paraphilia).

GENERAL CONSIDERATIONS IN THE CONSULTATION PROCESS

1. Diagnoses

If there was a major diagnosis (Axis I, according to the criteria of DSM-IIIR) such as an anxiety disorder or a depressive disorder, we would want to note that. In most instances, there was no such disorder. Homosexuality and adultery are not labeled as illnesses according to the criteria established by the American Psychiatric Association. One can make a good case for the diagnosis of Adjustment Disorder since almost all of the priests arrived reacting to a crisis that had been the immediate reason for the referral. It was imperative, however, to see whether there was a definite personality disorder (Axis II) and if not a full-blown disorder, what were the dominant personality traits. As part of the diagnosis we also wanted to appraise the level of functioning both vocationally and personally. I was also interested in whatever psychodynamic formulation could be reached on the basis of a few interviews.

2. Vocational Considerations

Almost always there was some conflict with either church authority or with some person or a group within the church, or, at least the potential for that was possible. What personality attributes accounted for the conflict? Given those personality traits, what recommendations could be made for work within the church?

Sometimes, those personality characteristics, given the situation, might exclude the clergyman from additional church work. At other times, a shift in job from pastor to some secular administrative post might make sense. In some instances, no change in the priest's role was recommended.

3. Recommendations for Future Therapy

Since this was, first and foremost a consultation, although in most instances it had a positive psychotherapeutic effect, there was concern for future treatment. If future treatment was indicated, and it usually was, what would be its nature? Outpatient psychotherapy or perhaps hospitalization? If psychotherapy, what about the nature of the psychotherapy recommended, i.e. dynamic, cognitive, behavior, or supportive? Might medication play a role?

A significant part of the appraisal revolved around the clergyman's concept of self. Here were men with a "calling" whose behavior had exceeded the norms of his church. Did this change the nature of that calling? What had happened to the priest's self doubts and conflict between his ego-ideal and reality? How did the priest square his conscience with the transgressions and their consequences? The inquiry into these aspects of the consultation process was intensified when Dr. Slowinski, who had been trained as a seminarian, joined in the consultation as he did in over half the cases reported in this paper.

4. Consultation Method

The priest would generally come and spend four or five days being interviewed. If married, when indicated, his wife would accompany him and be interviewed, as well. Some of the interviews were therapist and patient, therapist and spouse, others were conjoint couple interviews, and if Dr. Slowinski was participating he would duplicate the interviews, and he and I would carry out some of the interviews as a co-therapy team. His participation made it possible to compare notes with a colleague and to arrive at a joint appraisal and recommendations. It also enabled us to check our perceptions and discuss our counter-transference feelings to the priest to make sure that one's reactions were not idiosyncratic.

In addition to interviews, we used a battery of psychological tests. This included a marital problems check list (if indicated), the Beck and the Zung depression scales, the MMPI, the Millon Personality Inventory, and the 16 PF. In a few cases these were augmented by the Loyola Sentence Completion Blank for Clergymen, and the Multimodal Life History Questionnaire. In several instances, the priest had already seen a psychologist and we had the opportunity to review the psychologist's report prior to the consultation.

5. Confidentiality

Since the priest had committed a transgression, usually identified by telephone calls from Bishop Richards and often with the priest's superior, usually a bishop, prior to our seeing him, how could we establish an atmosphere in which the priest would feel safe? How could we provide a "holding" environment for someone who easily might perceive us as being in league with church authority whom he actually had offended or feared he might offend? The only way to achieve this, to provide for the clergyman a feeling of security and trust in the process, was for us to be completely honest and above-board. In the majority of instances, no report in writing to the bishop in authority was necessary. The priest knew we would discuss the matter with Bishop Richards but in almost all instances he had already talked with Bishop Richards and felt comfortable with that. However, if a report to a churchman in authority was indicated, then we would tell the priest that we were required to do that, but that we would share our report with him. The last session was a "wrap up" conjoint session in which

Dr. Slowinski and I discussed our findings and our recommendations with the priest so that he would know exactly what we were saying and to whom we were reporting. To make the consultation therapeutic and psychologically and emotionally helpful, an atmosphere of safety and genuine concern for the welfare of the priest had to take place.

Another aspect of confidentiality revolves around family secrets. If the priest/husband had a secret gay life or was involved in heterosexual affairs without the wife knowing it, or at least was keeping some aspects of his behavior from his wife, we impressed on the priest the need to share this information with his wife during the course of our consultation. Generally, the wife knew all or most of her husband's secrets by the time the two arrived in the office but there were several exceptions to this which had to be dealt with. If we felt that there was some secret behavior, the knowledge of which would have been devastating to the relationship, we might have hesitated in maintaining our insistence on "coming clean."

HOMOSEXUALITIES

As mentioned above, seven priests were referred because of known or suspected homosexual behavior. Of these, five had identified themselves as gay, although with varying degrees of certitude. One of the priests, age 53, claimed that the only same sex behavior that he had ever engaged in after college had occurred two years ago when he had, on several occasions, fellated a sixteen year old boy in his church chambers. As the priest said, "These are the only unethical acts that I have knowingly committed during the twenty-five years that I have been with the Episcopal Church." He had never thought of himself as homosexual.

His few homosexual contacts with a college roommate had been dismissed because his roommate was actively heterosexual and the homosexual acts had been limited to mutual masturbation. On several occasions in his life, he had been approached by other men, even a church superior, in a homosexual solicitation, but he had rebuffed the advances. The episode with the young boy came at a time when the priest had been impotent for six years because of diabetes and one of his sons had turned to drugs and was in a great deal of difficulty. The priest had an enormous need for affection which he was not getting from his wife or children. In his view, his action had been a loving albeit a misguided one. The young boy seemed to be a substitute for his own son.

What are we to make of isolated homosexual episodes such as this in a priest who was remarkable for his devotion to the church and who had done an excellent job, according to all the reports we received? That there was latent homosexuality in him would be difficult to deny, but, according to him, he had not even been aware of any erotic interest in males. Had these fantasies and impulses been so well suppressed during all these years only to break through in a moment of crisis? Is this a good example of latent homosexuality?

More typically, the priest had been trying to manage his homosexual impulses and fantasies, usually masturbatory fantasies, by suppression. In some instances, the onset of homosexuality occurred in mid-life after the death of the priest's mother or wife. The range of the onset of homosexuality in five of these priests was from 30 to 52 with an average age of 42. That is unusual and indicates the degree of suppression exerted. In only three of the seven was there a long history of gay behavior and these three men, two of whom were married, thought of themselves as definitely gay. This gay behavior of one of the married men had increased significantly after his wife's death, 12 years earlier. Both married men were certainly capable of heterosexual intercourse with some degree of enjoyment. Nonetheless, there was preferential homosexuality. Another priest, also married, whose case is described above certainly did not think of himself as gay through his 25 years of marriage. A single priest had never married, indeed, was still a virgin although in his fifties. He was asexual and the closest he came to open expression of his sexuality was in kissing a female or wrestling with a young man. It was the latter that had become suspect.

There were two priests whose gay behavior occurred after years of marriage and homosexual abstinence although it was clear that they had had homoerotic impulses from an early age. These men had only really faced their homosexuality in mid-life. The variation in gender identity and behavior is so great that the term "homosexualities," indicating the pluralism, used by Bell and Weinberg [2], and again by Sipe is preferred to "homosexuality."

THE CRISIS

Each priest arrived for consultation in a crisis situation. Something had happened to bring to the church authority the priest's alleged gay behavior. In the case of the priest who had fellated a young boy in his church chambers, the boy's lover who was a few years older, accused the priest and made sure the bishop knew of it. In that instance, it was a pure case of revenge. The lover had been refused admission for training in the church and angrily sought revenge against the priest and the church itself.

In another case, the priest had brought a young man, whom he had counseled for more than a year, back to his home to spend the night, after the young man had complained that he had nowhere to sleep.

The priest did this despite his knowledge that the young man came from a criminal family, that the young man's brother was in jail and a sister was a well-known prostitute. During the course of the evening, he asked the young man if he could get into bed with him, whereupon the young man attacked him savagely, breaking many bones in his face and beating him half to death. In discussing the situation with the priest, I learned that at an earlier time in the priest's life, in somewhat similar circumstances, he had refused to take in a homeless young man and the young man had committed suicide. The priest had suffered enormous remorse over this and his recent action was clearly a reflection of his wish to avoid a similar outcome. Unfortunately, his erotic impulses got the better of his otherwise good judgment.

One priest got into trouble when the police raided the public meeting place where he had been meeting other gay men. Once on the police blotter, his situation became widely known to the church and to many in the congregation.

Another priest, married for many years, and apparently quite capable of enjoying heterosexual intercourse, would spend vacations with another couple. The priest became sexually involved with his "good friend." Later, the other couple moved to another city.

It was the priest's misfortune that letters from his friend were sent to the wrong address and inadvertently opened. The person discovering the letter recognized that a priest was involved so he brought the letter to a friend of his, a member of the congregation. In this way, the priest's gay behavior became widely known throughout the church. One priest in his 60s had been homosexual all his life, although he had been married until his wife's death 12 years prior to the consultation. He was well known to the gay community in his area and, presumably, would have been able to keep his homosexual behavior discreet, if not for an attempt at blackmail by a young man. The blackmailer had been arrested and had received a fine, after which there was an attempt on the part of the blackmailer's friends to blacken the priest's reputation by accusing him of having sex with young boys. The priest denied that he had ever had sex with anyone under the age of 18.

PROBLEMS OF THE CHURCH WITH HOMOSEXUALITY

The church confronts several major issues concerning homosexuality. One is the attitude of the church toward gay parishioners, especially if they are asking for a gay marriage or wish to adopt children. Is homosexual behavior a sin or is it not? Should homosexuals be welcomed into the church on an equal footing with heterosexuals?

An even more troublesome problem, one dealt with in this paper, is the attitude of the church toward gay ministers. It is clear that a significant number of clergymen are gay in their self identity and probably at least half of these engage in gay behavior. Although the exact number is probably unknown, one can guess that it is at least twice the percentage found in the general population, an estimate made as well by Sipe about Catholic priests. If five to ten percent of men in general are gay, we can assume that at least 10 to 20 percent of clergy are gay. This is because religious life attracts a higher percentage of gay men, presumably somewhat more true of the Catholic church than of other denominations because of the protection that celibacy affords against the demands of a heterosexual life. Nevertheless, it probably is true that religious activities in general appeal to a higher percentage of gays than is found in the general population.

Presumably, if the gay minister can keep his activities discreet, he will be relatively safe. However, if he marries there is always the threat of his secret life being discovered by his mate as has happened in several of our cases, or some event brings the matter to the attention of the police, or some indiscretion, sometimes involving blackmail, makes the minister's private life a public one. Since there is a great deal of homophobia in the general community, the minister's homosexual activity is often not generally acceptable. This is particularly true in smaller communities.

The general attitude of the church is "If you must engage in gay behavior, keep it out of the limelight. Don't let it become a matter of public interest and speculation."

In all the cases reviewed here there was no attempt on the part of the church to cover up the activities of the priest; instead, he was referred for consultation and in all the cases therapy was recommended. The therapy was not designed to change the priest's homosexual behavior but to have him keep his behavior under stricter control so that he would not hurt himself or embarrass the church.

TYPES OF SEXUAL ABUSE

The behavior of the priests with alleged or actual gay behavior leads to the conclusion that there are two types of sexual abuse. One type involves the abuse of conventional morality; the second involves abuse of the pastoral role. It is important to differentiate between the two, since the second is a far graver form of offense, and strikes at the heart of the pastoral relationship.

The violations of conventional morality fall into two groups. In both, private behavior becomes public but sometimes the involvement of the police raises the possibility of criminal charges, if the priest is the accused, or adverse publicity if the victim. An example of the second group is that of the priest who had sex with his best male friend and their relationship was disclosed inadvertently by the interception of a letter. The first group (involving police) is exemplified by the priest who was picked up in a police raid on a well known gay meeting place.

Abuse of the pastoral role involves an abuse of the power of the priest or clergyman invested in him by the parishioner and the failure to maintain boundaries between his professional and personal roles. Any professional, such as a physician, has power and authority by virtue of his training and credentials, creating in the patient or client an expectation that the professional will be helpful, that his intentions are therapeutic, and that his self-interest will not interfere with his role as helper. The symbolic role of the clergyman as religious leader magnifies the power and authority of the professional role. The priest is a representative of the tradition and history of the church, of faith, and even of God. (It would be interesting to know if the public expectations of the minister, and the investment of power are about the same or different from the expectations of a judge. I believe for most people, they would be higher.)

Violations of boundaries between professional and personal roles augment the abuse. Just as a father must maintain certain boundaries with a child, so must a priest refrain from engaging in sexual activity with a parishioner; otherwise, it is very much like the incest that occurs within the family. Fortune states, "The Church is frequently referred to as the Family of God. Jesus referred to his followers as his siblings and to God as his parent. This image suggests the positive parallels of trust, intimacy, caring, commitment and respect which should be the bases of human family life" ([4], p. 83). This is yet another way in which the violation of role boundaries by a clergyman exceeds those of a physician or therapist.

Examples of such abuse in the group of gay priests are the priest who fellated a boy in the church chambers, the boy being one of his flock, and the priest who brought a young man to his home and wanted to climb into bed with him. We can understand the dynamics in each case, in the first the search for substitute love, and in the second the wish not to relive the guilt that followed the death of another young man, and hence we can sympathize, forgive and have compassion for these priests. Nonetheless, these were violations of the power of the priest and the boundaries of professional and personal roles for which the priest must be accountable.

HETEROSEXUAL ABUSE

With these distinctions in mind, let us turn to the five priests who were accused of heterosexual violations. Recall the two types of abuse, namely, the abuse of conventional morality and abuse of the pastoral role; all these priests abused their pastoral role to varying degrees. I will describe these five cases starting with one who had to give up his parish priesthood even though his abuse of his pastoral role was far less severe than that of the four other priests, three of whom were multiple offenders. This priest had an affair with one of his parishioners. His wife found out about it, left him, and the scandal forced him to resign his position. If the priest's mistress had not been a member of the congregation this could be listed in the first category and would not be considered an abuse of his pastoral role. But she was a member of the congregation; therefore, it had much more profound implications for the parishioners.

This case was unlike the others in one other respect. On the recommendation of Bishop Richards he came to me for therapy, not for consultation. His difficulty had led to his resignation prior to therapy. Since he had entered therapy voluntarily, the church did not need to stay involved. During the process of therapy, the priest came to realize that he had a conflict about his choice of career. He returned to his first love, teaching.

This made it unnecessary for the church to make any decisions concerning this man's future vocation. Based on the results of therapy, I feel reasonably secure that if I had recommended another church position, the church administrators would have accepted my recommendation. This case was unusual, in that the priest was deeply in love with his wife, and yet his family background made him vulnerable to the kinds of adulation a priest very frequently receives from his female parishioners. His father had been prominent in political circles and highly regarded as a very successful man. Growing up, the priest had never felt that he could live up to his father's expectations. He tried entering his father's profession but sabotaged himself academically, making it necessary to leave professional school. Joining the ministry was an attempt to find purity of heart instead of the rage and bitterness that he felt at not being able to secure his father's approval. The priest's mother was an alcoholic whose behavior, while being somewhat seductive, was also unpredictable, leaving the boy with feelings of distrust and uncertainty about her love. The priest's generally low self esteem was further eroded when his first wife deserted him for another man after two years of marriage. Her suggestion that they have an open marriage because she found the other man more sexually satisfying was a source of deep humiliation. Fortunately, soon after, he fell in love with his second wife.

At another church, in another locality, the seductive aspect of the clergyman's role and its power to seduce women became very apparent and it was then that he had his first of two extramarital relationships. His wife discovered the adulterous liaison, leading to a good deal of conflict and tension in their marriage, but she stayed with him until a few years later when the situation repeated itself in another church.

In therapy, he discovered why he was so vulnerable to the attention of women. It not only restored his diminished self esteem but it temporarily gave him the illusion of security. He had taken his wife's love for granted, especially when she stayed with him after the first of his extramarital relationships. He learned, too, how possessive and demanding he had been of his wife. Fearful of rejection, he had made unceasing demands on his wife for reassurance that she loved and cared for him. His excessive control of her had led to much conflict in their relationship, increasing his vulnerability to the attention of another woman.

His wife, who was still in love with her husband despite a gratifying relationship in the year or so since they had been separated, came in for marital therapy. The marriage was restored, although with a good deal of difficulty in the early months.

This case illustrates a milder degree of confusion between the professional and pastoral roles than the other cases to be described. In this instance, even if the priest's mistress had not been a member of the congregation, the chances are he still would have been forced to resign because of the violation of conventional morality. In the remaining cases, the abuse of the pastoral role is much more evident.

Three of the four remaining adulterers had a history of womanizing. They got into trouble when they chose women connected with the church. In two instances, the women themselves "blew the whistle"; in another the cuckolded husband of the priest's mistress called the relationship to the attention of church authorities. What is so striking is the long years that active womanizing can go on, without it coming to the attention of church superiors. All three were married, albeit unhappily.

Sometimes, bad luck combines with bad judgment. One of the priests had a long history of extramarital relationships before and after his ordination as a priest. His had been almost an unconsummated marriage for years as his wife suffered from dyspareunia and vaginismus. Eventually, they divorced and he fell in love with a woman he desired to marry. This was an unfortunate decision for two reasons; she was married and she was an official of the church. The rejected husband brought the matter to the attention of the church.

This is a priest with great gifts whose failure to maintain the boundaries between his professional and personal life has caused harm to himself, his mistress, and to the members of the congregation. His career still can be salvaged with appropriate psychotherapy.

The other two womanizers are more typical of the case described by Fortune in *Is Nothing Sacred* [4]. She describes the Reverend Dr. Peter Donovan of the "First Church of Newburg" as a seducer of dozens of women in the various congregations under his pastoral care. Finally, six women had the courage to present their complaints, eventually leading to the resignation of the pastor. The pain and anguish of the women, the denials of churchmen and their attempted cover-up with the consequent delays and procrastinations, and the imperfect resolution of the problem are the core elements of Fortune's book.

One of the priests had a roving assignment to a variety of missions and he brought to his assignments a roving eye. Repeated relationships finally caught up with him, as several woman connected with the missions finally complained of his unwanted attentions.

I would like to describe the remaining case of a womanizer in greater detail for it illustrates well the failure to maintain boundaries. One of the illustrative facets of the case is the great attraction felt by many women for a priest, even more so if the priest has a high rank. The seductiveness of power has been noted many times in history, so that even physically unattractive but powerful men have no shortage of adoring women. This is a case in point. This priest had had women throw themselves at him at every church assignment leading to a succession of many seductions and affairs.

The corruption of power is a two way street. The powerful man is himself seduced and misled by his power, warping his judgment. This priest took on the professional role of counselor to another priest and his wife, the very woman with whom he was sleeping. During the same period of time, he was having intercourse in various motels and other places of assignation with another married counselee and asking still another member of the congregation to provide cover for them so that the clergyman's wife would be kept in the dark. It is no wonder that this "house of cards" would soon collapse. Retribution was swift and resignation was the consequence.

What was the nature of a marriage that supported this arrangement for three decades? The wife was the child of an alcoholic father who was also a child molester, and although the wife denied being molested by her father, the evidence is substantial that she was not spared the pain and the humiliation of a sister and a cousin. She was programmed to be co-dependent.

What about the priest? His father was a brutal sadistic man who often physically attacked his mother. She would come in to the boy's room to hide from her drunk, abusive, and dangerous husband. This went on until the father died when the son was a senior in high school. His experiences left him with a mental script in which he was a substitute for father (husband) and protected and loved a helpless but dependent mother (wife). This led to a series of affairs, all with married women who were at the same time his professional clients or counselees.

Understanding his and her individual psychodynamics, as well as their conjoint marital dynamics *does not* exculpate his guilt. He is still accountable for his actions. He knew what his pastoral role was supposed to be; he knew what his profession demanded of him.

What are these men like? They are vain, dependent, histrionic and at the same time truly insecure. They need the attention and adulation of women. Often they have feminine attributes demonstrated on the MMPI. Women throwing themselves at them restore their sense of masculinity and simultaneously their sense of being loved. They feel as if they count for something. It is a true psychic "fix."

One case in which a priest was accused of adultery remains for discussion. This is a man who is not a true womanizer. This case represents a situation in which a marital pair have reinforcing and interlocking pathology. The behavior of the clergyman's wife is a powerful addition to the pastor's effect. Together they can cause quite a mess.

This priest and his wife were so overly nurturing to their parishioners that they forfeited a private life, individually and together. Their home was always open to members of the congregation at any hour of the day or night. For years they had taken vacations with three or four congregation couples, sharing a rented home at the shore or mountains. This enmeshment with the parishioners and with each other led to a series of blunders caused by poor judgment, involving allegations of misuse of money, bypassing vestry and church authority, all under the guise of being do-gooders. Eventually, the enmeshment led to an accusation of sexual impropriety, denied by the priest. Some of his letters to a married woman in the congregation would lead almost anyone to believe that it involved a sexual relationship. In interviewing the priest, I was not sure, but on the basis of clinical experience, I would have to assume that it did. The pathology of giving (in order to receive) excess love was so strong that it could have led, at the very least, to inappropriate language on the part of the priest.

Another priest whose difficulty, interestingly enough, was not sexual transgression put it well in one of his letters to his standing committee that was evaluating his behavior:

The other issue concerns my relationship with women. I hug and kiss both men and women with a Christian intention and understanding. Throughout my life women with passive, dependent, hostile, violent, or uncommunicative husbands have misunderstood the Christian love (agape) and understanding given by a priest. I have never known a priest that is loving that did not have rumors around them. That is the price if you live openly and do not draw back from or reject those around you. Quite frankly, I prefer the company of men because I like hard physical work and fishing. People who wish to spread rumors can make much out of this statement.

All the more reason why clergy of whatever denomination should recognize the need for circumspection. The priest, whom I quote above, was faulty in his judgment of the permissible, and his failure to understand the impact of his actions and words on others caused his difficulties within the church. This priest thought he was God's gift to women. He certainly had nothing good to say about their husbands. There is often a thin line between self-assurance and grandiosity.

Fortune calls for preventive training in seminaries. The seminarian needs "to understand the nature of the power and authority of their role and the responsibility that goes with it. They need to learn how to maintain boundaries and relationships with parishioners and counselees. They need to learn to care for their own emotional and sexual needs in appropriate ways" ([4], p. 56).

PARAPHILIAS

"Paraphilia" literally means "away from love" and does not carry the judgmental implication of "sexual deviance." In paraphiliac behavior, the individual has a compulsive need to substitute some other erotic activity for vaginal intercourse. Examples are fetishism, voyeurism, exhibitionism, frotteurism (rubbing, touching), sadomasochism, etc. The diagnosis can be made if the patient is markedly distressed by "recurrent intense sexual urges and sexually arousing fantasies of at least six months duration" (DSM-IIIR) involving the particular sexual deviation such as fetishism or pedophilia.

For the church, the most important paraphilia is pedophilia. The opportunity to sexually molest children is enhanced by parish duties, wherein priests have frequent contact with children of all ages. If a minister or priest is found to be a sexual molester of children the outrage is enormous.

The term pedophilia refers technically to the molestation of pre-pubertal children. More girls than boys are molested. Finkelhour [3], who questioned hundreds of college students about their sexual experiences in childhood, reported that 11 percent of the females and 4 percent of the males stated that they had had an unwanted sexual experience with an adult before the age 12. We do not have a satisfactory term for sexual activity with post pubertal children. In the case of a pedophiliac priest to be described below the psychologist to whom the priest was referred for treatment used the term "ephebophilic." That Greek term literally relates to young manhood, but this priest molested girls. The term does not quite fit but it may be better than anything else we now possess. This priest molested girls between the ages of ten and sixteen so that his victims might be call "parapubertal." There are men who are erotically turned on by girls prior to the development of secondary sex characteristics and there are also those who are turned on by girls as they develop breasts, their figures round out, and they have the beginning of pubic hair. My patient was one of the latter, and even though some of the girls were ten and eleven, in most instances their sexual development was already becoming apparent.

THE CASE OF A PEDOPHILIAC PRIEST

This man, in mid-life, was referred to me after a young girl in his congregation, 12 years of age, told her parents about his fondling of her. Her parents were outraged and protested to the church authorities. The priest did not deny the allegations and was suspended from his ministerial functions pending the outcome of consultation with me. He reported that he had been molesting children for more than twenty years. He estimated that there were between 50 and 60 girls all from the ages of 10 to 16 who he has touched, hugged, and caressed. Most of the time it was through their clothes, the primary attraction being their small budding breasts (frotteurism). On a very few occasions, when the relationship had lasted several months, he had stimulated their genitalia. In the process, he gets turned on and has some evidence of a seminal emission but claimed that he had never ejaculated.

He had had pedophiliac fantasies starting in his teen years but had not engaged in any relevant behavior until about the age of 30. After that he would always be able to anticipate situations in which he would be alone with young girls, as a consequence of his activity in the Girl Scouts and other church-related classes attended by young people.

About 12 years before consulting me, he had to leave another church because of complaints by the parents of two young girls. But in more than 20 years of these activities he had been in trouble on only these two occasions. He seems to have remarkably little remorse for any potential damage that he had caused these girls, unconcerned about the impact of his behavior on any of them. On a depression inventory he had checked off, "I don't feel particularly guilty," "I don't feel disappointed in myself," "I don't feel I am any worse than anybody else." Indeed, his score on the Beck Inventory was zero indicating a massive denial of psychological distress. As will be discussed below, this is a typical attitude for a pedophile.

The priest's wife had only been informed about the accusations 12 years earlier and of the current ones. She really had no idea of the frequency and longevity of her husband's contacts with young girls.

One of the remarkable psychological facets of this case was that despite the patient's "almost complete absence of introspective capacity" as reported on psychological tests, he talked in overt terms of his sexual attraction to his mother when a young man, especially his attraction to her breasts. At the same time, he was extraordinarily shy as a teenager and young man. He found it extremely difficult to approach a young woman for a date.

It might be speculated that his pedophilia was a defense against incestuous feelings for his mother as well as a defense against possible rejection. Given the power the priest had over these young, passive, and for the most part accepting young girls the pedophiliac activity provided a defense against rejection. When I pointed out the abuse of power, he seemed startled, as if he had never thought of his behavior in terms of power.

The priest was referred for behavioral therapy, with a very positive result. At the end of treatment he was no longer responding physiologically to arousal stimuli to which he had previously responded. As the therapist wrote, "Objective physiologic assessment demonstrates the elimination of ephebophilic patterns of arousal and the restoration of more healthy heterosexual patterns." The priest had relocated in a new setting and obtained non-church-related employment. What would be remarkable if it were not the typical pattern is the length of time the priest was engaging in these activities without detection. Even the prior accusation had not led to his removal from his ministerial duties. Unfortunately, this seems typical of paraphiliac activities of priests. The case of the Catholic priest in Louisiana, which became the focus of a TV documentary "Judgment," is another example.

In that case, the Catholic church had to pay out millions of dollars in damages and is unable to attain sexual liability insurance from any insurance carrier. In this case, the estimate by the priest that he had molested 50 to 60 children is not unusual. Reid (quoted by Sipe) wrote, "Child molesters commit an average of 60 offenses for every incident that comes to public attention" ([5], p. 85). On average, the child molester victimizes 76 children [1].

Catholic priests, according to Sipe, are different from Episcopal priests, or other ministers, in that three quarters of Catholic priest pedophiles are homosexual, whereas, in the general population, heterosexual abusers outnumber homosexual by 2 to 1. Sipe makes the additional relevant comment that the strong sexual attraction toward pre-pubescent children probably is already present in a large number of seminarians: "Because their clerical training is generally protective and does not challenge their psychosexual development, many of these men are unaware of the extent of their developmental deficits and sexual tendencies until after ordination. Few of them act out extensively prior to then" ([5], p. 45).

TRANSVESTISM

One of the priests referred was a transvestite. He had gone to his bishop and told him that he was resigning from the Church since he felt that his own sexual activity was incompatible with his own image of ministerial behavior. The

Bishop had asked us to see him in consultation and to recommend how the Church could assist this priest in helping him find some relief from his misery. The priest had just informed his wife that he was leaving her, as well as the Church. He had been married for three years and, although he valued the relationship, he felt it was not fair to his wife to continue a marriage that was beset with so many problems. There was no doubt that he was depressed. It was not only apparent clinically but on his Beck Inventory he scored a 25, confirming the presence of a moderate depression.

This priest's history was fairly typical of transvestites. He had been overprotected by his mother, sleeping with her nightly for several years while his father was hospitalized for a chronic illness. He had started to cross-dress at about the age of 10, about a year after he had exposed himself to a group of girls in the neighborhood. He still remembered the terrible guilt he had experienced following that episode. As a teenager and young adult he had been a member of a street gang and fighting with knives and guns had been frequent. Only family connections kept him out of jail.

Hyper-masculine behavior is typical of the transvestite. It is as if, with the wearing of female clothing, the person can be both male and female at the same time. The sexual excitement produced by clothing, especially tight female underclothing, and the erection it produces reinforced his masculinity while at the same moment he "gets under the skin" and possesses the female without running the risk of rejection or being hurt by a woman. All of these dynamics were present in this priest.

His reaction to his sexual compulsion, a paraphilia, however, was to feel terribly guilty and somewhat hopeless about the future. Our task as consultants was to steer him to the right therapist who could then help him sort out his life and help him find work that would not create the guilt which he had been feeling as a priest. Whether the marriage could still be salvaged would remain for the future. At the time of the consultation his wife was equally ambivalent about the relationship.

There are many transvestites who maintain their marriages. For the marriage to be successful the wife has to accept her husband's secret sexual life. If she can, the marriage may work out; if she finds that impossible, the marriage is doomed. In this case, the wife had known about it four months after their marriage and had stuck it out with her husband. It was he who felt too guilty to continue, believing that he was depriving his wife of a better relationship.

In many instances, cross-dressing of this sort (as opposed to transsexualism in which the male wishes to have his genitalia removed) is compatible with the role of a clergyman, unless, as with this man, he feels that his sexual life is incompatible with his image of himself as a priest.

SEXUAL ADDICTION

Most instances of sexual addiction are paraphiliac in nature. Occasionally, one finds addictive behavior in which there is not a clear-cut sexual deviation. Such was the case of a priest who, over a period of thirty years, had seduced more than 100 women. Most of the women were connected with the church. There was an interesting twist to these seductions in that most of the women consented to being photographed. He had 121 photographs of nude women in his collection, discovered by his wife when he left them in a paper bag in the garage. On several occasions he had attached some of the woman's pubic hair to the photograph making it in effect a fetish which he used to increase his excitement while masturbating. This behavior demonstrates that paraphiliac elements were not completely absent. Voyeurism and fetishism were minor motifs in the scenario of seduction.

Once again, as we have seen so often, for thirty years this priest had been "getting away" with his unministerial behavior. It finally came to the attention of his superiors when a woman in the congregation whom he was seducing, decided to marry a man of another religion, who was pursuing her at the same time. The rival discovered the liaison with the priest and reported it to the church authorities who demanded and received the priest's resignation.

The dominant theme in this case is power. The primary motivation for the priest's behavior was to achieve domination and power over women. He had been victimized by his mother, who was a sadistic tyrant, who kept a whip at her side during every dinner and would lash the youngster's legs if he misbehaved. She also treated her husband with enormous contempt and humiliated him at every turn.

The relationship with mother was reenacted with the priest's wife. He grew to hate her as much as he hated his mother. A chronic illness, which started several years after their marriage, was sufficient excuse for him to start his extramarital adventuring. The first of these occurred during his wife's first pregnancy. She quickly discovered her husband's womanizing. It appeared as if the husband was only too happy to let his wife know what he was doing, again, demonstrating his power over her. As happens in many marriages of clergymen, wives hang in because of the loss of income and status that would ensue if divorce took place.

Once again, this case is a rather extreme example of the abusive power of someone in an authoritarian role, especially if the authoritarian role coincides with the role of counselor, and of religious advisor. Also typical, unfortunately, was this man's almost complete lack of remorse for the pain and suffering he had caused.

SUMMARY

A series of cases have been presented demonstrating the abuse of power in clergymen who refuse to recognize or subscribe to boundaries between professional and personal roles leading to the victimization of parishioners or congregants.

The issues demanding attention are:

1. How seminaries can be helped to select trainees who are less likely to become sexual transgressors. How to provide early detection of clergy whose sexual development is seriously flawed and to provide treatment for them. How to modify curricula in seminaries so that these issues are faced fully and squarely during training.
2. How to achieve prompt recognition of sexual abuse of church members and to provide appropriate support and counseling for victims of clergy abuse.
3. How to provide appropriate therapy for clergy with fixed patterns of sexual transgression.
4. What are the ways in which the sexism of society infiltrates the various denominations, and reinforces the masculine traditions of church authority? Much of church history, tradition, and ritual surrounds the dominance of man over woman, even the perception that woman is the "embodiment of evil." How do these factors affect the problem of sexual transgression? How can the Church change its attitude and teaching?
5. What are the administrative obligations in response to the detection of clergy abuse? What should be done regarding the employment of clergy who have violated the ethical codes of their profession? In what circumstances is rehabilitation possible?
6. How can the churches adopt a code of ethics, a self-governing disciplinary code, depending on the peer system of evaluation and enforcement, modifying or even eliminating scriptural interpretation and hierarchical regulation?

The Center for Sexuality and Religion, recently incorporated and now undergoing development, may be a place where many of these issues can be thoroughly debated and policies recommended. (For information about the Center address: P.O. Box 945, South Orange NJ 07079-0945.)

University of Pennsylvania and Pennsylvania Hospital
Philadelphia, Pennsylvania, U.S.A.

NOTE

[1] For a full discussion of this issue see [5].

BIBLIOGRAPHY

1. Abel, G., Mittelman, M. and Becker, J.: 1987, 'Sexual Offenders: Results of Assessment and Recommendation for Treatment', in M.H. Ben-Aron, S.J. Hucker and C.D. Webster (eds.), *Clinical Criminology: The Assessment and Treatment of Criminal Behaviors*, Butterworth, Toronto, Canada, pp. 107–128.
2. Bell, A.P. and Weinberg, M.S.: 1978, *Homosexualities: A Study of Diversity among*

Men & Women, Simon and Schuster, New York.
3. Finkelhour, D.: 1979. 'Psychological, Cultural, and Family Factors in Incest and Family Sexual Abuse', *Journal of Marriage and Family Counseling* 4, 41–49.
4. Fortune, M. M.: 1989, *Is Nothing Sacred: When Sex Invades the Pastoral Relationship,* Harper and Row, San Francisco.
5. Sipe, A.W.R.: 1990, *A Secret World: Sexuality and the Search for Celibacy,* Brunner/Mazel, New York.

DAVID E. RICHARDS

ISSUES OF RELIGION, SEXUAL ADJUSTMENT, AND THE ROLE OF THE PASTORAL COUNSELOR

Among the care givers in today's society the pastoral counselor occupies a role that is considerably less clear than the role of psychologist, psychiatrist, social worker, or sex therapist. Historically the pastoral counselor is seen to be 'connected' in some way with organized religion and religious institutions, and it is often assumed that he bears responsibility for seeing to it that clients who come to him are guided along pathways that are in keeping with religious traditions and the customary ethical norms associated with religious practices – no matter how vaguely or inconsistently these may be defined.

The dictionary definition for "counseling" suggests the following: it is "an exchange of opinions and ideas"; it may be perceived as "consultation" or "discussion"; it is "advice or guidance, especially as solicited from a knowledgeable person"; it may be "a deliberate resolution", a "plan", or a "scheme." This way of thinking about counseling seems to infer that the care giving service of the pastoral counselor is essentially an action which answers questions, provides direction and may result in a deliberate resolution or plan. This conceptualization of pastoral counseling is reinforced by popular religious imagery – specifically so in the Christian tradition through reference to the term pastor or shepherd. A shepherd – usually presented as a male – has responsibility for watching over sheep who are generally pictured as quite helpless and completely in need of his guidance from pasture to pasture and his protection against threats in the environment. This pastoral figure is perceived as strong, and his task is to care for the weak. His work is to minister authoritatively to the needy and less informed. If he does this work well the population under his care is protected both from danger and from error. They may never learn to take care of themselves or to make their own decisions, but within certain limits it is likely that they will always be safe.

SEXUAL HEALTH AND PASTORAL CARE

This stereotypical view of the function of pastoral care giving is probably accurate in describing how things used to be and may even be accurate in reflecting how many religious adherents continue to feel about the pastoral care giving role. The fact is, however, that in some parts of the world of religion and among some adherents remarkable changes have taken place in the past 50 years. Before commenting on these changes it is relevant to our purposes in this publication to note that with this conceptualization of pastoral care giving we might guess that issues relating to sexuality and sexual ethics would have been

Ronald M. Green (ed.), Religion and Sexual Health, 187–202.

dealt with in a somewhat dogmatic manner with little opportunity for open exploration of fears and feelings and a rather heavy emphasis on certain rules and the correctness of behavior.

In describing such conditions we keep in mind that we are referring to a time in religious observance and practice when there was limited availability of sexual knowledge and information and when it may not have seemed appropriate for religious leaders, teachers and pastors to be concerned about the use of such knowledge and information as may have been available. The informed use of contemporary insights and the use of the significant body of knowledge which is constantly emerging has not been characteristic of religious bodies until the last two or three decades.

Today, however, we see commissions or committees on sexuality active in most major denominations. It is impressive to note that in its master plan for the future The Sex Information and Education Council of the United States (SIECUS) sees the churches as allies in the struggle to achieve effective education for sexual health throughout our nation. The recent SIECUS publication entitled "Sex Education 2000: A Call to Action" has this as an objective:

> *GOAL FOUR: By the Year of 2000, All Religious institutions serving youth will provide sexuality education.* Religious organizations have an important role to play in sexuality education of children and youth. These institutions are well suited to discuss values about sexuality – a subject that is often difficult to discuss in school or community programs. Religious congregations have traditionally determined community ethical standards, and many families turn to their religious institutions for guidance and counsel. Religious institutions can present children with information about sexual values that relate to the institutions' overall moral and ethical standards ([8], p. 9).

This clearly suggests that a major change has been underway in recent years in terms of the degree to which sexual health is seen as a concern of organized religion.

IMPACT OF CLINICAL TRAINING

As this change has gradually been occurring in the field of organized religion, significant changes have been taking place with regard to the role and function of the pastoral counselor. Perhaps the most important factor impinging on pastoral care and pastoral counseling and producing change has been the creation in the middle 1940s of a new educational resource for theological students which emerged originally under the title of Clinical Training. As this resource has developed it has come now to be called Clinical Pastoral Education.[1] In England a parallel development under the influence of highly motivated and deeply religious psychiatrist, Dr. Frank Lake, has come to be known as Clinical Theology.[2]

A clinical orientation in the training of theological students tended to move students and the practice of counseling away from the more authoritarian and dogmatic positions toward a more non-directive way of helping persons who came seeking guidance and support. Carl Rogers's emphasis on non-directive counseling was a major influence along this line. At the same time the movement included a fairly strong reference to Freudian psychology and tended to increase among many seminarians an understanding of and respect for psychoanalysis. Within this context issues regarding sexuality were dealt with in greater openness, and pastors could not help but become more aware of the sexual and ethical dilemmas that were now current in our society. This clinical movement brought pastors-in-training in contact with psychotherapists. It became somewhat common for persons attending seminaries – at this point almost entirely male – to enter therapy, and we can only surmise that in the therapeutic setting they were called upon to confront their own sexuality in ways not provided to them through traditional seminary life and experience.

The combination of clinical training and the availability of psychotherapy probably enabled many pastors and pastoral counselors to deal with issues of sexuality more comfortably and with greater knowledge than that possessed by pastors and counselors in the pre-clinical era. There is still no body of data or information that tells us how the seminaries themselves have adapted to the changes brought about through the introduction of clinical training for their students. Impressions gathered in observing seminary life today suggest that by and large seminaries still do not provide a great deal of help to students in understanding their own sexual health and in knowing what the pastoral role is in helping counselees, congregants, and adherents in regard to sexual problems and dysfunctions. The attention currently being given to the various incidents among clergy of inappropriate sexual acting out and sexual abuse of various kinds suggests that seminaries need to give more attention to the issue of sexual health. A first step in this direction would be a well planned and carefully designed study of seminary curricula to determine how much or how little attention is being paid to sex education, sexual issues, and sexual health among students in the various seminaries and schools of theology throughout the United States.

As we compare the pre-clinical type of seminary training with the post-clinical it seems obvious to suggest that counselors and pastors who are now less directive and less 'guidance' oriented probably approach sexual dilemmas in a quite different way than in the past. Issues are less simple and more complex. Answers do not come quickly and easily. The struggle for ethical correctness is more intense. The demands for compassionate support and encouragement in the face of deep personal pain and inner conflict are much greater. The Pauline injunction in Galatians 6:1–2 is perhaps now taken with somewhat greater seriousness and with the recognition of its relevance to the times in which we live: "If we live by the Spirit, let us also walk by the Spirit. Let us have no self-conceit, no provoking of one another, no envy of one another. Brethren, if a man is overtaken in any trespass, you who are spiritual should restore him in a spirit of gentleness. Look to yourselves, less you too be tempted. Bear one another's burdens, and so fulfill the law of Christ."

Among the newer insights that have come to the field of pastoral care is the notion that the effectiveness of counseling depends on how accurately the counselor understands the state or condition of the person seeking help. In other words, a careful diagnosis must precede the commencement of the helping ministry. This classifies counseling as something more than temporarily relieving the anxious or lovingly comforting the distressed. Emotional and spiritual band-aids are no longer in style or acceptable. The pastor needs more than a generous heart, time to listen, and a modest endowment of common sense. He needs special knowledges and skills that help him discern with accuracy exactly what is being presented to him. Without these he has little choice but to fall back into the older role of general comforter or authoritative guide and director to dependent and troubled persons who are needy and unquestioning.

SEXUAL ISSUES IN COUNSELING

When called upon to counsel with persons regarding religion and their sexual adjustment and in seeking first of all to make an accurate diagnosis the pastor must be prepared to raise for her/his own consideration a number of relevant questions. Among these are such questions such as:

1. What does it mean when a counselee volunteers extensive information about sexual activity and behaviors and seems to be compelled to be totally forthcoming regarding these matters?

A clergyman in mid-life came for pastoral counseling after experiencing job difficulties that arose as a result of an episode of sexual acting out with a woman in his congregation. As the counseling proceeded it became evident that he had a need to talk about his sexuality. He was unusually open and forthcoming. The details of the affair were discussed and then he went on to reveal that over the years, as best he could remember, he had had about 25 sexual partners other than his wife. On an earlier occasion he had divulged information of this nature to a spiritual director. He had frequently resolved to get things under control. No success had been achieved along this line. He had come to recognize that he was in deep and serious trouble so far as his sexual health and sexual adjustment were concerned. The Carnes Sexual Addiction Scale was administered. The results indicated that sexual addiction was a possibility for him. The Sexual Dysfunction Unit at a nearby hospital was consulted. The client and the pastoral counselor agreed with the hospital staff that residential treatment was indicated. Because the issues around sexual adjustment had emerged so quickly and were so clear, the decision to enter the hospital program was made within three days of the beginning of counseling.

What does it mean when a counselee avoids totally all reference to sexual activity, behaviors and feelings even when there is evidence that there are severe problems in this area?

The Rev. Mr. Cox had over the years been suspected of inappropriate advances of a sexual nature to younger women. Finally, on one occasion the person whom he approached in this manner accused him of harassment and of being sexually abusive. Even though he persistently denied all accusations, he agreed to see a counselor. While he complied initially with the requirement that he seek counseling, he did not continue in counseling. Before too long another accusation was made. He continued to deny and refused to deal with the matter in any way. His absolute reluctance to face what was demonstrably a sexuality problem in his life placed him at risk in his parish and community ministry. A) The injury to other persons, B) the likely possibility of damage to himself, and C) the danger of scandal lead his supervising official to require his immediate resignation with no prospect of reemployment.

There are times when in the face of avoidance and denial such strict measures are required.

How can one help the couple who come seeking help who are unable to talk about their sexual adjustment when they come together but invariably raise sexual issues when each has the opportunity to see the counselor alone?

Mr. and Mrs. Jay came to seek help regarding their marriage of 27 years. They were parents of three children who were grown and out of the home. Both spouses were now employed. The nest was empty and the coffers fuller than ever before. However, they had adopted a pattern that kept each of them extremely busy and out of communication with one another. As they came together they could only focus on the mundane issues of housekeeping, their petty criticisms of one another, and the general frustrations they felt with life. In an individual counseling session he raised the issue of lesbian tendencies in his wife. In separate counseling she expressed alarm at what she suspected to be her husband's emotional attachment to another woman. It seemed that in pastoral counseling they could not face issues of sexual adjustment together. They were referred to a psychiatrist who was family systems oriented with the hope that ultimately they would work together on their sexual adjustment with someone who was new for each of them and with whom they had not had any kind of previous relationship.

This illustration raises another question that the pastoral counselor must consider: At what point – if ever – is it appropriate to raise directly with clients, and particularly with those clients with whom one may have had an on-going relationship as a result of normal church affiliation and pastoral interaction, specific questions about sexual adjustment? Is this area usually to be avoided and referral made to a family therapist or sex therapist? Must the pastor somehow limit his interactions to issues of religion and other aspects of personal life and allow some other care giver to deal with sexual adjustment? Or can the pastoral counselor be expected to deal with both of these areas?

Bob Clark sought counseling to talk about the dysfunctionality of his wife Jane. She had had many problems over the years. There was hardly any area of her life in which he felt she was competent and functioning well. Bob and Jane were seen together and separately for several months. Each continued to be intensely critical of the other. Bob was reserved and uncommunicative. Jane was alternately angry and depressed. On one occasion one of them made a passing and fleeting reference to dissatisfaction in their sexual relationship. However, the continuing dialogue focused on the logistics of housekeeping, money, and tension within the family unit. The counselor felt that the contacts that he was having with Bob and Jane in the community and through congregational activities were making it difficult to address this sensitive area. The decision was made to refer them to The Menninger Institute for outpatient marital evaluation with the hope that with clinicians who were entirely unconnected to them and in a more distant clinical setting they would find it easier to confront the sexual aspect of their marriage.

Responding to these questions regarding sexuality and the counseling role will test counselors' diagnostic skills and clinical judgment. The response will also depend on the extent to which she/he has been trained to deal with issues of sexual health. With adequate training and with comfort in regard to one's own sexual adjustment some counselors will be able to deal in many instances with both religion and sexual adjustment with confidence and effectiveness. However, a counselor's position within an ecclesiastical system will sometimes make this difficult if not impossible.

MISCONDUCT AMONG CAREGIVERS

While pastors and pastoral counselors face the test of their skills and their judgment it has become more apparent than ever before that they also face certain dangers that pose a threat to their careers. Recently the Rev. Marie Fortune[3] in her book *Is Nothing Sacred* has documented the increasing awareness to sexual misconduct among clergy and pastors. The media has also contributed a great deal to this mounting awareness by the attention they have paid to certain sensational incidents about which the whole nation is now quite well informed. Coincidental with this public awareness is the attention now being given by church authorities and judicatory officials to the formulation of ethical codes for clergy and pastoral counselors. In many church systems the mechanics are now in place for handling allegations and grievances and designating disciplinary procedures for offenders. As a part of this new awareness a great deal of concern and responsibility is being directed toward the appropriate care of and ministry to the victims that are involved. It is encouraging to note that in many instances this entire problem area is being seen as a systems-wide concern and responsibility. Movement in this direction promises to bring to the field of pastoral care and counseling a higher level of professionalism and the establishment of criteria and boundaries which will provide protection to both practitioners and clients.

DEFINING PASTORING

The term 'pastoral care' or 'pastoral work' or even the reference to being a 'pastor' carries with it a certain vagueness. It is not uncommon for church members to claim that they want a good 'pastor' without being entirely clear about what good pastoring really is.

Following is a paradigm for pastoral care that will help to delineate in broad terms the major categories that fall within the area of pastoral care giving.

Level I – Situational Crisis

- Illness, death, accident, sudden change, trauma.
- Requires prompt personal attention by pastor.
- Short term crisis-intervention ministry needed.
- May require longer term follow-up ministry.
- Issue is helping people survive stress and pain.

Level II – Crisis of inner conflict

- Mental illness, marital conflict, sexual acting out, addictions, personal dysfunctionality of various kinds.
- Requires willingness to self-disclose.
- Therapeutic intervention usually needed.
- Referral, following accurate diagnosis, to some other care giver may be necessary.
- Issue is dealing with pathology and dysfunction.

Level III – Growth and the use of life

- Deepening understanding of faith/belief system, study, learning, use of educational resources, spiritual guidance and direction.
- Requires training, teaching, coaching.
- Skill in helping people to grow is essential.
- Goal is to help all adherents reach their individual and appropriate vocations, and spiritual fulfillment.
- Issue is developmental and preventive and pertains to the use of life.

During the period we have designated as the pre-clinical era the general view of pastoral work focused largely on Level I. Pastors were the natural persons to call in quickly when tragedy struck. They were most commonly pictured in the sick room or at the death bed and often first on the scene of an accident or some other traumatizing event. At Level II their interventions were largely supportive and directive. Level III was covered by regular preaching, by teaching Bible Classes and classes for initiates and new members. This larger issue of helping people identify their gifts and genuinely helping people find their way in life and the work of reinforcing faith and belief systems as a means of preventing inner conflict and its accompanying manifestations of pathology were generally not seen as pastoral functions. Churches often employed a lay Director of Christian Education to cover the need for some of this teaching and training.

One of the interesting features of contemporary religion is the way in which this paradigm is now being reversed. Assisting with situational crisis as in Level I is still an on-going responsibility. As indicated above many pastors are now more competent and more expert in dealing sensitively with Level II and issues that have to do with deep inner, personal conflict. What is new is the claim now being made on pastors for dealing with the need for intellectual and spiritual growth among their constituencies. With regard to religion and sexual adjustment this moves the preoccupation away from a focus on sexual rules, violations of understood norms, and sexual dysfunctions. The movement is now more expansive and toward sexual health and constructive ways of integrating sexuality with the whole of life and particularly with one's spirituality.

SPIRITUAL DIRECTION IN PASTORAL CARE

Spiritual direction, which is clearly a part of Level III pastoral work, has, within the past decade or so, risen to a position of very great importance. The historic role of spiritual director has always been a part of the Roman Catholic tradition and has been most commonly associated with the contribution that the members of religious orders have made to the faithful laity and secular clergy of the Roman Catholic tradition. The retreat houses conducted by Jesuits and Dominicans and Franciscans and other orders have been for centuries havens of spiritual refreshment and sources of spiritual direction to pious individuals who have taken religious faith and practice seriously. Today, however, many others are learning the language and the vocabulary and the skills required for effective and meaningful spiritual direction and formation. Spiritual Life Centers of every denomination are appearing in many locations. Continuing education courses and special seminars on spirituality for clergy are multiplying. The contemporary literature on spirituality is expanding rapidly. In the 1950s I happened to know of only of one periodical, entitled 'The Spiritual Life' published by The Descalced Carmelites, which dealt very specifically not only with prayer and meditation but with the entire range of subjects included under the term 'spirituality.' Today there are a remarkable number of such periodicals – many published by religious groups but a number published by secular groups for whom the area of spirituality has become a major concern.

REFLECTIONS ON SPIRITUALITY

As a term, spirituality is frequently used without much definition and with the assumption that generally speaking everyone knows what we are talking about. Under these circumstances there is at least the possibility that the term means quite different things to different people. Since in any exploration of the theme of religion and sexual adjustment this critical element of spirituality cannot be ignored it may be helpful at this point to comment briefly on an understanding of spirituality which seems germane and relevant to this exploration.

While volumes could be written, and certainly have been written on this subject my comments will be limited to four brief statements. First, attention given to spirituality is in the interest of achieving the total integration of life at the deepest possible level of one's being. Within the religious framework this means making an effort to see that all of life comes under the direction and the influence of God. This effort is in response to human beings' understanding of their own finitude. This awareness is reflected in such things as the Alcoholic Anonymous member's absolute reliance on a Higher Power for health, sanity and salvation. This same need was identified and articulated in many ways long before the days of AA, but this is one of the most dramatic contemporary acknowlegements of a person's need for the total integration of his or her life around some powerful center. Pascal puts it in this way in his *Pensées* (Sect. VII, 425) when he states that in every person there is "the infinite abyss (that) can only be filled by an infinite and immutable object, that is to say, only by God himself." For the religiously-minded person, having an authentic spiritual life means an active commitment to the idea that no aspect of life is outside of or apart from God's will for one's life.

Second, spirituality is the means by which we develop, with a strong sense of consistency, the idea of intentionality about our lives. In these days many corporations and organizations of all kinds make much of writing their "mission statement." This is nothing more than a corporate statement of intentionality. I suspect that many of the persons who devote considerable energy to composing the corporate mission statement have not yet written their own personal mission statement. The writing of a personal mission statement should be a deeply spiritual experience because it is one's highly individualized statement of how she or he intends to make use of the gift of a human life. Being clear about one's purpose in the conduct of one's life is a major step toward being a well differentiated self and is a primary objective in one's spiritual pilgrimage. Understanding the role that sexuality will play in one's life and arriving at that understanding with intentionality on our part is an important aspect of that pilgrimage.

How sexuality relates to life's spiritual pilgrimage is a question that is sparsely covered in the literature of spiritual formation and spiritual direction. I suspect that historically among spiritual directors and counselors the issue of sexuality has rarely been approached unless it has become a problem area due to either sexual acting out or to neurotic preoccupation with sexual themes.

Sexuality generally has only become an issue for discussion when something seems to be going wrong.

Parents, teachers, pastors, doctors, and others in positions of influence would seem to prefer to make grand assumptions of satisfactory sexual health as children are growing up until danger signs appear. As adults these assumptions continue, and sexuality does not come up for consideration unless something is off the track.

The mind-set that produces these conditions is based on the primacy of sexual roles rather than the primacy of values that are inherent in human sexuality. The subliminal message from birth onward is simply "don't break the rules. Observe the moral precepts no matter what and everything will be all right." This makes it difficult for spiritual pilgrims to raise questions regarding their own sexuality. A discerning person might wish to ask: "What do I really do with my sexuality? How do I constructively and creatively relate to the sexuality of other persons? To what extent must I regulate my sexual behaviors? To what extent is it right for society (including family, religious groups, employers, peers, etc.) to regulate my sexual behaviors? Am I always at the mercy of my biological urges? Where does my life in spirit intersect with my sexuality? How do I balance restraint (i.e. self-denial) and expression (i.e. sexual behaviors)? How does my own conceptualization of appropriate sexual behaviors impact other areas of my life? Apart from what I may think of as 'good' or 'bad' how do I come to understand what is beneficial for me and what is harmful? How shall I intend to use and enjoy my sexuality? And above all, how will I integrate it with the other essential aspects of my life such as love, productivity, and relating to others?"

Responding to these questions and discussing options and alternative ways of thinking that might be available would perhaps help a person to become more intentional regarding his/her sexuality. Getting consciously clear about our intention for the use of the totality of our life is neither easy nor the result of magical or instant insight. It is the fruit of spiritual growth and development taking place over time.

Third, a spiritual life is shallow or perhaps non-existent if it does not embrace a commitment to an organized effort in some form. It is not made up of good but vague thoughts. In the Christian tradition the term Rule of Life refers to the organized effort by which an individual seeks to gain guidance, support, and divine encouragement for continuing the effort. The term "Rule of Life" is an established one in the field of ascetic theology (i.e. the theology of the spiritual life). It is based upon two assumptions. The first has to do with the omnipotence and the goodness of God. She/He is understood to be unfailingly consistent. The second assumption is that human beings find it very difficult to be so consistent. Therefore, in making an effort to grow religiously and spiritually it is important to submit to a discipline. The term asceticism is derived from the Greek word 'askesis' which means 'exercise' or 'training'. A Rule of Life is a self-imposed set of regulations which when followed will lead to spiritual growth. This is the point at which specific reference is made to prayer and meditation; although the life of prayer – no matter how active and regular – cannot be the limit nor the total expression of spirituality. It is simply an important means for carrying forward the on-going spiritual pilgrimage.

Fourth, spirituality is the arena in our lives in which we formulate, select, define, articulate, modulate and develop a value system. Values and spirituality are closely associated. Thinking within a religious context we generally believe that our spiritual life dictates our values, and that the absence of a significant spiritual life will also shape or form of one's value system – in whatever limited way that may be expressed. This brings our focus, then, to the issue of values and sexual health [6].

VALUES AND SEXUAL HEALTH

Values are the motivational forces that drive our lives. They do not come about within us by accident or by sheer luck. There is a design by which they are formed at the very center or core of our being.

For some people this process of value formation is intentional and constitutes a primary function of their spirituality. Values are selected, nurtured, cultivated and brought to a stage of fruition as one aspect of one's spiritual growth and maturity. But values may also emerge in people's lives simply as a result of developmental influences that have dominated or directed their behaviors – influences such as the family of origin, the culture into which they are born, or the religious institutions that contribute to the nature or type of their inner journeys. It is, in fact, possible for such influences to be so powerful in an individual's development that there may be no intentional selection of value patterns. What then emerges, within that person, is a clear set of held and practiced values, but values that are not held nor practiced within a context of reflection and ongoing values formation.

Unfortunately, only in this context is it possible for us to reach and grow through adulthood with value systems that are entirely appropriate for each level of our development. Although our value systems begin largely as a part of our natural endowment, they have the potential for growth, change and development as life proceeds.

Several decades ago, the term "values clarification" entered our vocabulary. The term itself suggests that values are never absent. In fact, the inner, motivational forces that drive our lives always exist, even though we may choose not to name them. Values clarification, then, is a self-conscious, self-examining process of self-discovery that helps to discern the content of the inner motivational forces that make us who we are. Another way of stating this is to say that values clarification is a process of self-examination, self-evaluation, self-regulation, self-correction and finally self-affirmation, understood by some as maturity, integration or inner peace. No matter what words are chosen to describe the result of this process, it is the more or less ideal condition of being conflict-free and content with the degree to which our inner aspirations match our external behaviors. It is, for the most part, a state of being satisfied with the level of consistency achieved between our elevated ethical and other-regarding values and the way in which we live our life.

The delicate balance between highly-regarded values and behaviors is rarely a guaranteed, unalterable, steady state. It can be influenced mightily by other people, situations, the state of one's spiritual and physical health and by critical socialization factors such as marriage, friendship networks and the interactive patterns created by employment. At times such factors encourage us to advance in our values formation; but, regrettably, they also may cause us to regress.

VALUES AND RULES

How we behave sexually can be the result of values self-consciously selected and cultivated or of values that we appear to select but which are actually determined by the patterns of behavior that we literally "fall into." Allowing ourselves to "fall into" behavioral patters is, of course, in itself a values choice – one occurring in a non-reflective context. It is one that avoids the above cycle of self-examination, self-evaluation, self-regulation, self-correction and self-affirmation.

From a very early stage in life people seek guidance in ascertaining acceptable rules for governing their sexual behaviors. Teenage agony often runs long and deep with regard to this struggle; the parental terror that one's children will "break the rules" and "get into trouble" is a constant in many cultures throughout the world.

Over the years efforts have been made in some quarters to address this agony and terror by means of informed and effective sexuality education; however, there has often been resistance. This resistance has sometimes taken the form of passive, uninvolved indifference; but it has also taken the form of active criticism and outspoken opposition. Whenever and wherever resistance has been expressed, whether passively or actively, there has been the fear that someone was planning to tamper with the "rules" – rules that may not have been clearly agreed upon. Some people in our society would rather leave these same unclear "rules" undelineated and unclarified, in the hope that by doing so such vague "rules" will remain firm, fixed and unalterable. Unfortunately the long story of human sexual history suggests that a non-reflective and uninformed mentality leads more to trouble than to sexual health, wholeness and happiness.

One result of this particular approach has been the "just say 'no'"mentality. If we look for a moment at the value content of this "just say 'no'" posture we see a whole series of lower level values clustered around it. Primarily, however, there is a devaluing of knowledge and information. Informed judging and decision-making based on accurate information generally bespeaks a respect for human beings and the unique worth of every person. When absent, that which takes its place is the assumption that some persons know better that others how life should be lived. This assumption leads to the type of human stratification where those at the top of the hierarchy dictate to others what their values and rules should be. Learning and the reflective processes are devalued in this situation. In their place, rigid and dogmatic authority is accepted and highly valued and that which might foster growth is denied. This results in keeping people dependent and non-reflective.

This position, taken to be sure that no one will tamper with the rules, results in frequent, sharp and painful disagreements – disagreements that keep people so separated into opposing camps that no dialogue is possible.

A SYSTEM OF VALUES

One way to restore dialogue when it has thus been terminated is to move the discussion from rules to values. It is possible that if some degree of agreement can be achieved in regard to the values related to sexual health it may be easier to discuss the "rules" of sexual health.

Raising the question of the values inherent in any major life issue like sexuality involves one in an appraisal of one's entire system of values. For example, if one is consistent in one's values it is not possible to demonstrate for world peace while at the same time exploiting one's sexual partner. If one truly believes in the dignity and worth of all created beings then one cannot take sexual pleasure in total disregard for the feelings, needs and desires of the partner who shares in that experience.

Three steps are essential in exploring the values of sexual health. *First,* and fundamentally, one must self-consciously examine one's total value system. One of the earliest insights arrived at through values clarification is the understanding that a value is not a value unless it is acted upon. Verbalizations do not count; behaviors do. Therefore one's value system is made clear by assessing one's behaviors, by examining one's use of time, by looking at one's use of money and by watching every indicator in one's life that reveals, in some outward fashion, the inner motivational factors that drive it.

The *second* step is to state as clearly as possible what is understood by the term sexual health and then name the values that are associated with this concept.

In the third step one must compare ones values associated with sexual health with one's behaviors. The bottom line consideration in that comparison is whether or not in one's heart of hearts, in one's inner being, one genuinely feels that the things that one does in regard to one's sexuality do, in fact, embrace and act out the values that one believes in and respects. If there is that happy sense of consistency and congruence in this area of our lives then much of the time we will be at peace. If we feel in conflict, this is a good opportunity for more reflection and further growth.

SUMMARY

In this essay I have tried to address the issues of religion and sexual adjustment from the perspective of pastoral care giving and the role of the pastoral counselor. In recent decades certain changes have occurred in our perception of the role of the pastoral counselor and the work of pastors. However, seminaries or the appropriate denominational agencies responsible for training persons for the pastoral ministry have to look more carefully at the need that may exist for more adequate resources for helping pastors to deal with issues in their own sexual adjustment. This would, in turn, help pastors to deal more effectively with matters pertaining to sexuality in the lives of their congregants.[4]

There seems now to be in process a move to view pastoral work from a wider perspective. Whereas in days gone by pastors were relied upon largely to provide comfort, relief, guidance and support in times of crisis or conflict, there is now an increasing expectation that pastors will serve the pastoral needs of their people by assisting them to grow, by fostering the kind of strength and insight that may even help to prevent some of life's traumas, and most especially by responding to the current hunger for spiritual guidance and formation. The connection between spirituality and sexuality must be seen as clear and valid if the men and women of our time are to discover and enjoy that sense of wholeness which we believe to have been an original and integral part of the endowment bestowed upon us by God's loving act of creation.

The following quotation is taken from the "Quarterly of Contemporary Spirituality" and points in the direction of wholeness and health to which we are guided by creation-centered spirituality:

> We live in a time when we are affirming the goodness of creation. Moving away from the Platonistic dualism that influenced our Christian spirituality for so long, we are retrieving a healthier and more biblical notion of our own humanity as a unity of flesh and spirit. We no longer view salvation in terms of liberation from the flesh. Instead, it is precisely through our flesh that we live our faith and work out our relationship to God, ourselves, and other people, as well as to all of creation.
>
> Happily, in an emerging creation-centered spirituality, we can accentuate the positive dimension of our corporeality and come to a deeper appreciation of our dependence on and responsibility toward the material world. One fine expression of this is the growing concern for environmental issues and the recognition of our responsibility for our Mother Earth.
>
> Not only do we reaffirm the goodness of creation, but we also stress the dignity of the human person. We speak today of the importance of the self – its wholeness and its actualization. We seek to become authentically human, embracing all that this entails ([7], p. 20).

In seeking to become authentically human we are called upon to embrace our sexuality as a God-given aspect of our humanity and to do this in ways that are healthy and in harmony with our spiritual life and religious commitment.

NOTES

[1] The Directory of the Association for Pastoral Education [1] offers the following description of Clinical Pastoral Education:

Clinical Pastoral Education (CPE) was conceived by Dr. Richard C. Cabot as a method of learning pastoral practice in a clinical setting under supervision. The Reverend Anton Boisen enlarged the concept to include a case study method of theological inquiry – a study of the 'living human document.' As CPE developed, other CPE leaders expanded CPE to integrate into pastoral practice a knowledge of medicine, psychology, and other behavioral sciences. Today many supervisors emphasize the importance of pastoral relationships being formed through an integration of personal history, behavioral theory and method, and spiritual development.

CPE is theological and professional education for ministry. In CPE theological students, ordained clergy, members of religious orders, and qualified laypeople minister to people in crisis situations while being supervised. Out of intense involvement with supervisors, other students, people in crisis, and other professionals CPE students are challenged to improve the quality of their pastoral relationships. Through pastoral practice, written case studies and verbatims, individual supervision, seminar participation, and relevant reading students are encouraged to develop genuine caring pastoral relationships. Through viewing complicated life situations from different viewpoints students are able to gain new insights and understandings about the human situation. Theological reflection is important in CPE as pastoral people seek ways to integrate theology with life experience.

Essential elements in CPE include an accredited CPE center ready to receive students, certified CPE supervisor(s) to provide pastoral supervision, a small group of peers engaged in a common learning experience, providing pastoral care to people in crisis, detailed reporting of pastoral practice, a specified time period, and an individual learning contract.

[2] An account of the development of Clinical Theology in England is found in [5], the recently published biography of the movement's founder.

[3] Marie Fortune is the founder and executive director of Seattle's Center for the Prevention of Sexual and Domestic Violence. An ordained minister in the United Church of Christ, she lectures widely and serves as a consultant for churches where sexual abuse by the clergy has occurred. She is the author of *Keeping the Faith and Sexual Violence* [4], [3].

Her latest book, *Is Nothing Sacred* [2] is the story of a pastor, of six women who risked a great deal to come forward and tell the secrets of their relationship with him, of the many players who knew or did nothing or who knew and did something, and of a congregation that was broken apart by the pastor's actions. *Is Nothing Sacred* is designed as a blue-print for churches hoping to find effective solutions for sexual misconduct against parishioners by clergy and seeking to prevent its occurrence.

[4] The Center for Sexuality and Religion has been formed in order to create resources that could be made available to clergy, churches, seminaries and schools of theology. Following is a statement of the vision and mission for the Center and a brief description of one of its functions in serving theologians and religious professionals:

THE CENTER FOR SEXUALITY AND RELIGION VISION

The vision of The Center for Sexuality and Religion is to explore and enhance the relationship between science and religion on the issue of sexuality and to promote sexual health and sexual justice.

MISSION

The center for Sexuality and Religion is an interreligious and interdisciplinary entity dedicated to the following principles, goals, functions(*), and objectives.

(*) *To foster the competence and integrity of religious leaders in matters of sexuality.* Extensive research is available regarding sexuality in many professional fields, e.g. medicine, genetics, the behavioral sciences, sociology and anthropology. Theologians and religious professionals often fail to receive the benefits of such research. A channel for communication and dialogue needs to be created to assure the circulation of such information to those within religious systems. Familiarity with this information and resources will assist those within religious systems to address any sexually destructive behavior by religious professionals, as well as to provide a basis upon which to deal with the deeper institutional and theological issues involved.

BIBLIOGRAPHY

1. Directory of the Association for Pastoral Education, Inc., 1989–90. Association for Pastoral Education, New York.
2. Fortune, M.M.: 1989, *Is Nothing Sacred: When Sex Invades the Pastoral Relationship,* Harper and Row, San Francisco.
3. Fortune, M.M.: 1983, *Sexual Violence,* Pilgrim Press, New York.
4. Fortune, M.M.: 1987, *Keeping the Faith.* Pilgrim Press, New York.
5. Peters, J.: 1989: *Frank Lake: The Man and His Work,* Darton, Longman and Todd, London.
6. Richards, D. E.: 1990, 'Values and Sexual Health', *SIECUS Report* 18, 1–2.
7. Rucsil, E.: 1991, '"I Lost Myself, and Was Found": John of the Cross for Today', *Spiritual Life* 37 (1), pp. 18–24.
8. Sex Information and Education Council of the U.S.: 1989, *Sex Education 2000: A Call to Action, Sex Information and Education Council of the United States, A Report on The Current Status and Future of Sexuality Education for Children and Youth in the United States,* Sex Information and Education Council of the U.S., New York.

WILLIAM R. STAYTON

CONFLICTS IN CRISIS: EFFECTS OF RELIGIOUS BELIEF SYSTEMS ON SEXUAL HEALTH

Two areas in life where persons can experience conflict are in their sexuality and in their religious belief system. In the experience of sexuality, the conflict can be between one's sexual value system and one's feelings and behavior. In one's religious belief system, the conflict can be between what one has been taught to believe and what one's internal voice is saying. Here I propose to examine what happens when these conflictual areas are in crisis with one another.

These are exciting times. In the last thirty years, we have learned more about human sexual behavior and human sexual response than we have ever known. Almost all of our new knowledge is in conflict with traditional religious belief systems. Thus our religious institutions have been challenged to their core in order to be helpful and relevant to the sexual concerns of our time. In this essay, I will consider the frontier areas of our sexual understandings and look at how religion can help or hinder a person in his/her journey toward sexual health.

RELIGIOUS BELIEF SYSTEMS

Regarding sexuality, theological belief systems ask the question, "What does God (or nature) intend?" There are basically three types of religious belief systems in Judaic and Christian thought which affect attitudes and behaviors about sex. The first, often thought of as the traditional value system, judges specific acts of sex as holding moral or immoral value. The Act-centered theology behind this value system is both Judaic and Christian and centers itself around the following directive about sexual behavior. "So God created man in his own image; in the image of God he created him; male and female he created them. God blessed them and said to them, 'Be fruitful and increase, fill the earth and subdue it ...'" (Genesis 1:27–28). The Roman Catholic Church developed a belief system based on this scripture and made it Canon Law in 1918 ([4], pp. 18–31). This belief system can be graphed in the following way:

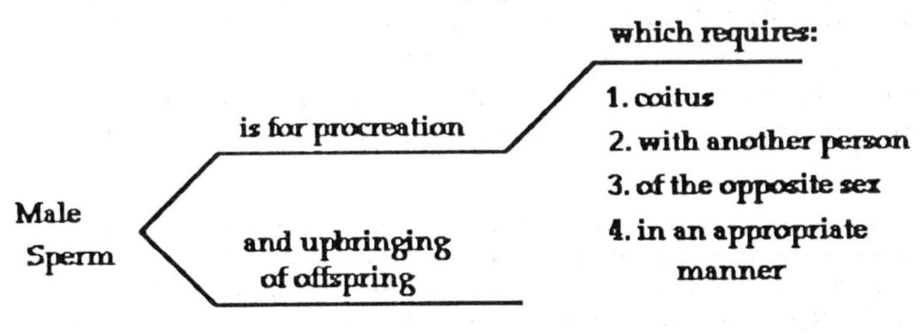

ACT CENTERED THEOLOGY

Ronald M. Green (ed.), Religion and Sexual Health, 203–218.

This view as it relates to sex, holds that the key to understanding God's (or nature's) intention for sexuality is male sperm. Male sperm is the basis of life itself. It is through the male sperm that life continues on to the next generation. The purpose of female sexuality is to be the ground in which the sperm (or seed) is planted. There is no concept that the ovum also carries the basis of life. Thus, in this theology, life is passed on through the male. The role of the female is to be supporter and nurturer of the male and his sperm. It does not matter what science or research says, these roles are fixed. This view has had enormous political and economic implications throughout history.

Male sperm, in this view, has two purposes. The first is to procreate. Sperm has no other functional use. One does not urinate better because of sperm, nor does it keep one from illness. The second purpose is more subtle, but just as important. Because of sperm there must also be a framework for the upbringing and education of offspring to pass life on to the future generations, otherwise it would be irresponsible to just run around planting one's sperm all over creation. There is then, the need for marriage and the family. This provides a responsible context for the development of the sperm.

While the Canon Law of 1918 seems simple and logical, it also implies what is not "allowable." For example, if sperm is only for procreation, then birth control and abortion are prohibited. Since procreation requires coitus, masturbation is certainly not allowed. Coitus with another person would make the fetishes or bestiality prohibited. Coitus with another person of the opposite sex prohibits homosexuality or bisexuality. And finally, coitus with another person of the opposite sex "in an appropriate manner," has been defined differently throughout the ages. It often includes disallowing practices like oral sex, anal sex, or even coitus in any exotic position. Parts of the church ordained the "missionary position" (or man on top) as being the only position for sexual intercourse that God would approve. After all, one plants seed downward, not upward or sideways. Even sexual intimacy in marriage for pleasure or fun was questioned.

Since sperm also needs to be within the context of marriage and the family for the upbringing and education of offspring, practices of premarital sex, extramarital sex, or any of the alternative sexual lifestyles would be prohibited and deemed "sinful" in this Act-centered system.

The acts themselves, described above, carry the concept of moral and immoral. Pope John Paul II is a good example of the purest form of this sexual theology. He regularly condemns each of the above "unallowables." He implies that it does not matter what science or research says about these practices, they are inherently immoral and even evil. A news report of a letter to the bishops of the church states:

> [H]omosexual tendencies ... are 'ordered toward an intrinsic moral evil, and thus the inclination itself must be seen as an objective disorder' ... The Vatican letter said that the church position 'cannot be revised by pressure from civil legislation or trend of the movement' ... [B]ishops must state clearly that homosexuality is immoral and resist pressure from the pro-homosexual movement within the church to change its teaching ... [B]ishops

'should keep as their uppermost concern the responsibility to defend and promote family life' (Philadelphia Inquirer, 10/31/86, 1-A).

At the other end of the theological spectrum would be a sexual value system based on relationships [4]. The answer to the theological question, "What does God (or nature) intend regarding sexuality?" would be that human sexuality is intended for relationships. A Relationship-centered theology says that there is nothing inherently immoral or evil about acts of sex. The important issue regarding sex are the motives and consequences of the act(s). This theological belief system is just as ancient, just as "biblical," has just as many "moral" spokespersons throughout history as the Act-centered theology. The Bible, in Relationship-centered theology, is seen as the history of human beings' relationship to self, to others (both male and female), to God, and to one's possessions and resources. The central focus of the biblical message was well stated by Jesus when he was asked what the most important commandment was in the law. He responded by quoting from Leviticus and Deuteronomy.

'Love the Lord your God with all your heart, with all your soul, with all your mind.' That is the greatest commandment. It comes first. The second is like it: 'Love your neighbor as yourself.' Everything in the Law and the prophets hangs on these two commandments (Matthew 22:34–40).

For Relationship-centered theology, to try to make the Bible say something specific about an act of sex is to misread the intent of God and the biblical message. The responsibility of religion, therefore, is to help individuals, families, and society to develop criteria for decision-making regarding sexual matters, taking into account the motives and consequences of sexual acts. The purpose of our engaging in sexual acts is to help us to grow as individuals, couples, families, and indeed, even as a society. The reward for this kind of decision-making regarding sexual acts is that life itself is lived more meaningfully in the present, as opposed to rewards in life after death. A good spokesperson for this theology was Pope John XXIII. He opened the windows of the Roman Catholic Church and started a movement that is still strongly felt in our world. In the area of sexuality, he suggested taking a new look at the issues around sex. An example of this movement was an excellent textbook *Human Sexuality: New Directions in American Catholic Thought,* edited by Kosnik et al., published in 1977. This study, commissioned by The Catholic Theological Society of America, questioned many, if not most, of the "unallowables."

Unfortunately most people are somewhere between these two theological systems, which becomes the third religious belief system. The third, or middle, belief system says that the Act-centered theology is what must be proclaimed and taught, but that when it comes to one's most private and personal decision-making system, a Relationship-centered theology is more relevant and "right." This creates conflict because these two systems are not compatible. An example of this conflict was demonstrated by a family (father, mother, and 13 year old daughter) who made an appointment to see me several years ago. The father told the following story.

Their daughter who had just turned 13 was pregnant. It was against the family's religious belief system, an Act-centered system, to have an abortion; but as they talked, searched their hearts, and prayed about the matter, they finally made a decision to have an abortion. He said several factors influenced their decision. The first consideration was the daughter's age. Second, she was a brilliant student and had a promising academic future. Third, she did not know who the father of the child was because she had sex with five boys at the same time. The family felt going through the pregnancy and having the child would deeply hurt her future. The father then told me that the decision was even more difficult and complicated because he was head of the "Right-To-Life" group in his community. He said that he believed in the views of that organization with all his heart and he wanted to continue in his leadership capacity, but that when it came to making the decision regarding their daughter, the "Right-To-Life" (Act-centered) value system was just not relevant to their personal circumstance.

As hypocritical as this family may sound, my heart went out to them because they were caught in a conflict between two incompatible religious value systems, an Act-centered versus a Relationship-centered theology. Proponents of each value system believe that they are right, moral, biblical, and that the other is wrong, immoral and unbiblical. While this was a very dramatic example of the conflict, I see individuals every day who are caught in the same type of conflict between an Act-centered theology and Relationship-centered theology. These theologies are not compatible. As hard as people try, the two sexual theologies cannot be reconciled or integrated.

AREAS OF SEXUAL CONFLICT

With these sexual theologies in mind, I will now turn to the types of sexual concerns that persons bring into therapy. There are five basic areas today of sexual concerns: sexual identity, sex-coded role, sexual orientation, sexual lifestyle, and sexual response and function. All these areas are controversial because they are closely tied to moral value systems. Whether religion is a help or hindrance to a person's resolution of problems in the above areas will largely depend upon which theological system the person adheres to.

Sexual Identity

Sexual identity has to do with being male or female. This may have at one time been a simple issue, but it is a very complex issue today. Being male or female is more than having a penis or vagina. There are a whole series of developments in embryonic and fetal development, from chromosomal to gonadal to hormonal to internal genitalia to external genitalia. Problems can occur all along the developmental pathway [8].

The mother always gives an X sex-determining chromosome. The father can give either an X or a Y chromosome. If both give an X sex-determining chromosome (XX), the child will be a female. If the father gives a Y sex-determining chromosome (XY), the child will be a male. But nature does not always keep things simple. There are many different types of chromosomal anomalies. It is possible that the father does not give any sex-determining chromosome (XO), and the child will be a female but often has distinguishing characteristics: she will be very short (probably not over 5 feet), may have webbing between the toes or fingers, and will be infertile. This condition is known as Turner's Syndrome. One out of every 2500 live female births is a Turner's Syndrome.

The chromosomal pattern could be XXY (Klinefelter's Syndrome). The child will be a male, but may also have some differences. He will often be tall, infertile, and may have difficulty in establishing good, close relationships.

The child could have XYY chromosomes. Some studies show that this person may have difficulty with conscience formation. A significant number of prison recidivists are XYY.

There are many more chromosomal patterns which can affect development and behavior. Several years ago, I had a family referred to me because it was thought that the young 12 year old son was a transsexual. He had started to cross dress when he was four years old and did it every day. I noted two things about him. One was that he was very short and had not yet started through pubertal changes. The other was that he had very few friends and this had always been true. Over a period of about one and a half years, he went through a variety of options: transsexualism, transvestism, homosexuality, and heterosexuality only to find that he was none and all. I finally requested a chromosomal test for him and discovered he had an XYXO mosaic. He will probably be vulnerable to all the sexual differences and lifestyles. This condition is now known as Noonan's Syndrome. His pastor and church were not helpful to the family at all, because everything the boy went through was considered immoral and sinful and not within the church's value system. The family felt lost and alone; their only tie to religious acceptance was through me.

Not only can there be different chromosomal patterns, but other differences can occur during fetal development. For example, in an otherwise female fetus there can be be an overdose of the male hormone, androgen. Instead of a clitoris, the child develops a penis and the family believes they have a boy, until the beginning of puberty, when he develops breasts and some breakthrough bleeding in the scrotal area, which turns out to be the beginning of menses. At this point the only thing the family can do is to have the child have a hysterectomy and mastectomy and to give some more male hormone to aid in the development of male secondary sex characteristics.

There can also be varieties of androgen insensitivity in male fetal development; that is, the fetus may have XY chromosomes, but when the internal and external genitalia begin to develop, an insensitivity to the male hormone causes the fetus to develop as female. The child is reared as a female, but because of her XY chromosomes, she will never menstruate or get pregnant. She may need some estrogen for female secondary sex characteristics and she may have to have surgery to lengthen her vagina. Otherwise she will be quite normal.

The question is what is a male and what is a female. There is no simple answer. When families have children born with these anomalies, whether their religious belief systems will be helpful depends on whether their religious belief system is an Act-centered or Relationship-centered system. The Act-centered theology will most often look at the parents as the cause of the child's difference and will not be helpful to their psychological adjustment to difference.

JEHOVAH, the Lord,... punishes sons and grandsons to the third and fourth generation for the iniquity of their fathers (Exodus 34:7).

A Relationship-centered theology would want to know more about the condition and to understand it, so that one's faith community could be more supportive of the person in relation to self, family and future.

Sex-Coded Role

Generally by the age of 2 or 2 1/2, everyone goes through a process of coding masculine and feminine attitudes, feelings and behavior [9]. Those attitudes, feelings and behaviors that one codes as appropriate for their own sex will code positive (+). Those that one codes as appropriate for the other sex will code negative (–). Using a male as an example, various types of coding are as follows:

IDENTIFICATION (+)
COMPLEMENTATION (–)

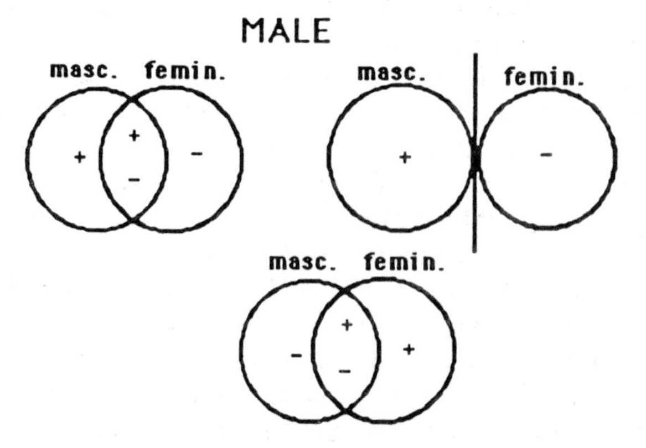

With the male as an example (although it could be just as well a female), we can see from the illustration on the top left the way coding takes generally takes place. For example, let us say that this male codes washing dishes as negative. This means that throughout life he will have to go against his nature to wash dishes, even though he may be a dishwasher by vocation. It will never feel natural to him. Let us say, however, that he codes washing dishes as positive (+), this means that he will always want to do dishes because it will be natural to him. He will not even have to be asked to do dishes. But let us say that he codes washing dishes in the ambiguous area. This means that if he marries a feminist who feels that housework should be equally divided, he will do fine with washing dishes. On the other hand, if he marries a traditionalist who never wants her husband to put his hands in dishwater, he will do fine in that relationship also.

The illustration on the top right is the man who codes both masculine and feminine positive, but there is no area of ambiguity. This is a transvestite. All transvestites are cross dressers, but not all cross dressers are transvestites. A transvestite generally codes relaxation and/or eroticism as feminine, thus to relax or to be erotic he must be in his feminine mode. Once coding has taken place, by ages two or two and a half, it is not likely that a person will ever be able to change. Therefore it is best for the transvestite and his partner to adjust to this fact, otherwise they'll fight it all their life.

Several years ago I had a client who was a transvestite. He had a Ph.D. in one of the sciences and held an important position in the aerospace industry. He had been married for many years and he and his wife had pre-pubertal twins, a boy and a girl. His wife had known of his need to crossdress since shortly after their marriage. His concern was whether and/or when to tell his children about his crossdressing. He decided to tell his children before they became adolescents. They have adjusted well to the information and, in fact, his daughter often goes clothes shopping with him. The children are now 16 years old. The only request his son has made is that his father not dress in front of his friends.

My client was curious to know if other transvestites would like to get together in a support community. I recommended that he inquire around and offered the use of my office for a meeting. He placed an ad in the local paper and there were 13 at the first meeting. This was four years ago. Today they have an average attendance of 100 at their once a month meeting and a mailing list of almost 700 in the greater Philadelphia area. Their meetings consist of a special speaker on some subject related to crossdressing, a support group for recovering alcoholics, a wives' support group, and a transsexual support group. I have been asked to speak several times and am always moved by the spiritual concerns of the group. None of them perceive their church or synagogue as being open to learning about the phenomenon of transvestism and transsexualism. Most of them feel that it is their deep religious faith that has kept them alive through all the trauma of growing up different from their friends. This client spoke at my church one Sunday and he wept through most of the service because he felt so accepted. He desperately wanted and needed to be a part of a faith community.

The bottom illustration is of a transsexual. This person, for some reason, (and we do not always know the reason; there are several good theories) codes opposite to their anatomy. The classical statement is for the person to say that they have always known that they were born into the wrong body. For the person who is brought up in a religious household, he/she will often pray from very early on that he/she will wake up the next morning in the body of the other sex. It is as if God perpetrated a cruel hoax by having them in the wrong body. It is best for this person to undergo sexual reassignment surgery because it is easier to alter the person's body than it is to change their mind to correspond to their body. In fact, I once had a transsexual client who went through 20+ years of psychoanalysis and in the end still felt he was a female. At age 63 this patient finally had reassignment surgery. She had three wives before the surgery and certainly tried to change. She was a Quaker and came from a Quaker Meeting that had a Relationship-centered theological stance. She believed that the only thing that brought her through surgery alive was her Quaker faith and support. When she died, I attended her Quaker funeral and it was very moving to hear testimony after testimony of how much she had meant to the life of her Quaker meeting as she went through her gender transition.

It is easy to see that the religious organization that follows an Act-centered theology would have great trouble with coding differences, thus making psychological adjustment at best difficult, and at worst, impossible. Act-centered theology would say it is not right to tamper with God's creation.

> No woman shall wear an article of man's clothing, nor shall a man put on woman's dress; for those who do these things are abominable to the Lord your God (Deuteronomy 22:5).

Transsexuals who come from an Act-centered theology have to supersede the mandate not to tamper with what God gave them to seek out gender reassignment surgery. I am constantly amazed at how many transsexuals have a very deep faith in God. While they often come from an Act-centered theology, they have felt deep in their heart that there was a loving God who accepted them and wanted them to be whole as a person in the other sex. It is this faith that carried them through troubling years as a child, adolescent, and then adult.

Relationship-centered theology would take into account what the medical community has learned in scientific research about gender development and gender roles and would do its best to help the person make a healthy adjustment to his/her gender role regardless of gender anatomy.

Sexual Orientation

One of the most conflictual and misunderstood areas of human sexuality is sexual orientation. Sexual orientation is defined as whatever it is that causes a person to have an erotic response. It is a complex issue. To begin with, the therapist needs to distinguish sexual orientation from behavior and identification. Sexual behavior is what a person does genitally with their erotic response. Sexual identification is how a person names themself, for example, as heterosexual, bisexual or homosexual. A person can be totally heterosexual in their behavior and identify as a heterosexual person and yet be predominantly homosexual in orientation. Such a person does not experience sexual health. Sexual health is being congruent in all three areas of behavior, identification and sexual orientation.

Sexual orientation is fixed early in life, probably no later than six or seven years of age and many researchers believe that sexual orientation is set much earlier [10]. We do not know all the factors involved in why a person is either homosexual or heterosexual. The majority of researchers today believe that we are born with a bisexual potential and in some way, either by nature or nurture (or both), get programmed either predominantly heterosexual or predominantly homosexual. This means that, depending upon the circumstances, the majority of persons could cross over from one orientation to the other, given the right set of factors. For example, it is common for both heterosexual males and females to engage in same sex behavior in a same sex private school setting or in a prison and yet, when out of that environment, to return to their heterosexual preference. Many homosexual persons have successful sexual experiences heterosexually, only to end up living an exclusively gay lifestyle.

Sexual orientation is more complex, however, than whether a person is heterosexual, homosexual or bisexual. Most people "turn on" (experience an erotic response) to a variety of things. My guess is that even the most repressed person feels sexually stimulated by several different things, while the majority of the population can "turn on" to many more things. The more common things in this category are certain foods, the ocean, music, certain smells, the outdoors, strenuous and pleasant physical activity, good conversation, candlelight, a mountain stream, a sports car, a particular color, explicit and/or romantic films, stories, certain rituals, nice clothes, and on and on and on. In fact there is a saying in the human sexuality field that there is something for everyone. I have attempted a paradigm regarding this theory, which has been labeled Stayton's Paneroticism [3], [10], [13]. In this model, I have attempted to show that people have erotic responses in all the dimensions of their relational life. A person can have an erotic response to himself/herself, so that masturbation can be a primary erotic expression. People also turn on to things in their life. One only has to sit in my office for a few hours in a day to find that there is practically nothing in this universe that someone does not turn on to, both in the animate and in the inanimate world. As a therapist who works often with religiously oriented people, I find that many people, especially deeply spiritual people, have an erotic response to God and to things of the Spirit.

We live, however, in a culture that tries to program us away from being sexual in any dimension of our relationships. Our religious institutions, especially those dominated by an Act-centered theology, are often the chief culprit in trying to program their constituents away from being sexual for more than procreational purposes. There is certainly little support for having erotic differences or being a member of an erotic minority.

The majority of things that persons have erotic responses to are harmless. For example, I have met a number of people who have an erotic response to rubber. Interestingly, a number of them are members of the clergy.

Several years ago I was speaking at an agency in another state and I stayed with a clergy family. I had one dinner meal with this family. Besides myself, there was the father (a minister), his wife, and their daughter who was a senior in high school. During dinner, the daughter asked me if I had seen her Dad's rubber room yet. With some embarrassment (regarding the question) I said, "No." She then told me that her father was a rubber fetishist and she got up from the table and brought over a magazine of the MacIntosh Society, which is the international rubber fetishist organization. (It was a very well done and expensive looking magazine. It was filled with stories and pictures of rubber artifacts and people in scuba diving suits.) After dinner, the family took me upstairs and showed me a room that B.F. Goodrich would have had an orgasm in. There were rubber artifacts from all over the world, from small intricate items to a large truck tire. His wife told me he was a great lover as they slept on a rubber sheet. I learned more that weekend than I ever imparted to my audience at the agency.

This is only one example of well over a hundred that I could have given. Sexual orientation seems like such a simple issue, yet there is more grief among individuals and couples about having erotic interests that are different. The Act-centered religious institution or clergyperson will be intolerant of these differences even though research may show that it is harmless and even necessary to the psychological and sexual wellbeing of the person. The religious institutions which proclaim a Relationship-centered theology need to become better educated about sexuality so that they can be more affirming of difference.

Sexual Lifestyle

Pastoral Counselors and most marriage counselors are trained to treat one sexual lifestyle – a strictly monogamous marriage. Depending on the counselor's theological position, other lifestyles may be tolerated, but seldom encouraged. Yet the overwhelming evidence seems to point to the fact that human beings are not monogamous by nature and that a strictly monogamous nuclear marriage is a minority lifestyle in this culture and indeed in every culture that has ever existed [2], [12]. In my work as a marital therapist, I have identified 14 types of relationship lifestyles that people are living in today. Only one of these lifestyles is a strictly monogamous nuclear marriage.

Dr. Helen Fisher [2], chief researcher for the American Museum of Natural History, studied both animal and anthropological research and concluded that for the past 5 million years, animals and humans (and their pre-human counterparts) have "pair bonded," ended the pair bonding on an average of four years, pair bonded again, and maybe again, and all the while practiced clandestine adultery. So our high divorce rate is not unique at all, but the variety of lifestyles that we have created may be a sign of our uniqueness. It certainly seems that those entrusted with doing our therapy and guiding our moral concepts and decision-making should be made more aware of human nature and not try to fit everyone into the same mold.

The lifestyle options that people are choosing today, beside traditional monogamy, are child-free relationships, single parenthood (as a choice, not because of a divorce), singlehood, cohabitation, serial monogamy, communal living, swinging and/or group sex, group marriage, synergamous relationship (where one has more than one committed relationship on a one to one basis), open-ended relationship (sometimes referred to as open marriage), chaste marriage (no sexual contact), family network system, and the secret extramarital relationship [12].

Most of the people that I see in therapy are in one of these alternative lifestyles (that is, alternative to a strictly monogamous nuclear marriage). I could give examples from each of the above categories. These people are often active in our churches or synagogues, yet seldom do they get either support for or help from their faith community for the lifestyle they have chosen. I very frequently get telephone calls from clergy asking me for advice on how to help a parishioner they have discovered in an alternative lifestyle. They feel their training has not been helpful to them in these real life situations in their congregations. I find, generally, that clergy are eager to learn and to change their rigid Act-centered theology. They are like the "right to life" family I discussed earlier in the theological center, who are trying to hold in some creative tension an Act-centered theology and a Relationship-centered theology. It just does not work, and these clergy often find they preach and teach an Act-centered theology, while practicing on a very private level, either in their own homelife or in their pastoral counseling, a Relationship-centered theology.

Sexual Response and Function

The final area of concern and conflict that individuals bring into the therapist's office involves issues around sexual response and function. Until the 1960s very little was understood about human sexual response. Masters and Johnson first described the sexual response cycle as developing through four phases: excitement (arousal), plateau, orgasm, and resolution [6].

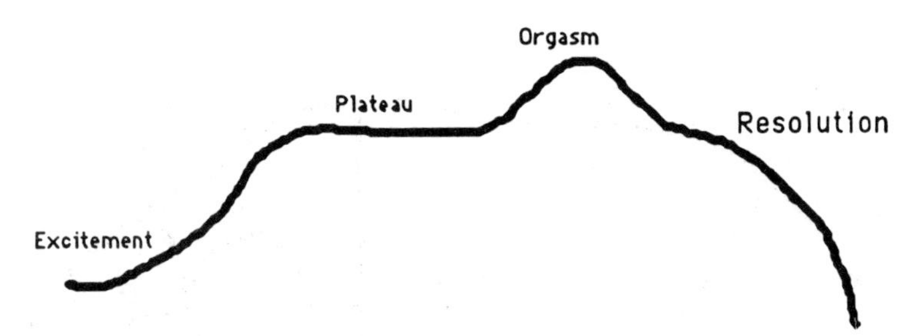

Masters and Johnson, 4 Phases of Sexual Response

Masters and Johnson actually measured, in both the male and female, what happens throughout the sexual response cycle genitally and non-genitally. They described the dysfunctions in terms of how the physiological response breaks down in reaction to disease, anxiety, trauma, medication, or changes in body chemistry [7]. They provided the foundation for research that has been conducted over the past 25 years. The important message from their research is that sexual response is a natural function, like breathing. It is not something that is added on at adolescence and taken away with the aging process. It is not limited to marriage or pair-bonding. It is in the very nature of life to have sexual responsivity. They discovered, however, that sexual response is the only natural function that can be inhibited for as long as a lifetime. In their book on sexual dysfunction, Masters and Johnson also found that a chief culprit in creating sexual dysfunction was "religious orthodoxy," which is based on an Act-centered theology. This was of such concern to Masters and Johnson that they sponsored a conference in the early 1970s to explore the impact of religious belief systems on sexual function and dysfunction. Representatives from all the major faith groups and from several human sexuality programs from all over the country spent a week in St. Louis considering the issue of the religious impact on sexual function and dysfunction. The result was a book *Sexuality and Human Values,* edited by Dr. Mary Calderone [1], and a resolve on the part of a number of the denominations to develop a better sex education program in their curriculum [11]. The implementation of the recommendations from this conference have yet to be realized, mostly because of the influence of the Act-centered positions of many of our denominations.

Another contribution to our understanding of human sexual response came in the 1970s with the publication of Helen Singer Kaplan's book, *Disorders of Sexual Desire* [5]. Dr. Kaplan made an important contribution to our understanding of the sexual response cycle. In her book, she describes all the ways in which sexual desire gets inhibited: from medication, to problems in the relationship, to sexual abuse, to the effects of disease and illness. She then outlined the therapy for sexual desire applicable to the various therapeutic modalities.

Another important contribution to our understanding of sexual function was the Erotic Stimulus Pathway theory of Dr. David M. Reed [3], [13].

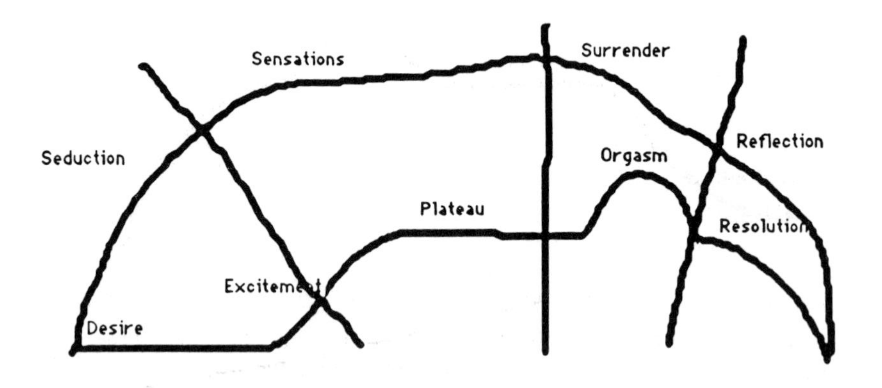

David M. Reed Erotic Stimulus Pathway

The Erotic Stimulus Pathway occurs in four phases corresponding to the theories of Masters and Johnson and Helen Singer Kaplan. The first phase is the Seduction phase. The art of seduction is both learning how to make oneself feel attractive and how to attract someone else. The most natural time to learn this art of seduction is during adolescence when a person is moving from the family as the primary relationship to the peer group as the primary relationship. The seduction phase becomes a repository of the many things that will eventually cause a person to experience sexual response.

As the adolescent gets older these seductive pleasures help to initiate sexual desire and excitement (or arousal). The sensation phase then gets activated as the senses begin to play a part in the sexual arousal. The senses of touch, vision, smell, hearing, and taste are nature's aphrodisiacs. These aphrodisiacs help to keep the person at a high level of sexual arousal during the plateau phase of the sexual response cycle.

Unfortunately, few, if any, sex education programs help young people to know the importance of their senses to sexual pleasure. In fact, the senses are often not related to sexual pleasure except in a negative way.

A couple was referred to me by their pastor because they had sexual relations only once or twice a year and the wife was very frustrated. The couple loved each other very much and were very committed to their marriage. When I took a sexual history from the husband, I asked him which senses brought him pleasure in lovemaking. Did he like to touch his wife's body or to have her touch his body? He said, "I don't mind if she rubs my back or my belly, but generally my skin is sensitive and I don't get much into touch." I asked him if he liked to see his wife nude or to watch when they were having sex? He answered, "I have never seen my wife nude. Bodies are gross. We get undressed in separate rooms and put on nightclothes, then we get into bed and, if we are going to make love, we turn the lights out, because watching sex is a real turn-off." I then asked him if he liked hearing the sounds of sex, which sexually arouses many people. He replied, "Frankly, I get turned off if I even hear her breathing. I prefer that she just lie there quietly, because if she begins to breathe heavily or make noises, I completely shut down." I then asked if the smell of sex was at all a turn on and he said, "Heavens no, she smells like a fish factory." With only one sense to go, I said "then I imagine that the taste of sex is out too" and he replied, "You better believe it."

The prognosis for this man getting over his sexual apathy was zero. When I asked him what he got out of sex, he told me that he had two pleasures in sex. First, he liked the feeling after orgasm. Second, he was "off the hook" for having sex for at least another six months. While this is an extreme case, I think it is an important illustration of how important our senses are to sexual pleasure and yet there is very little information imparted to people about their sexual response cycle and the factors that help us to function the way nature intended.

The third phase of the erotic stimulus pathway is surrender. Orgasm takes place when a person surrenders over to the experience. The dynamic of surrender centers on power and control. If there are control or power issues in the person or between the couple, then the orgasm phase will most likely be affected.

Finally, the erotic stimulus pathway ends with the Reflection phase. In many ways, this is also a very important part of the cycle. How the person reflects on the experience immediately afterward will become a feedback system that will affect the next sexual experience. For example, if a person reflects positively immediately after the experience, then he/she will look forward to the next time. On the other hand, if it was a bad experience or the person did not like the way he/she experienced himself/herself in the sexual encounter, then the reflection will be negative and it will have a negative effect on the next sexual experience. For this reason, it is important for a person's first experiences to be positive since this will build towards sex being a pleasurable and fulfilling event. If a person's first sexual experience is a result of child sexual abuse or rape, it then becomes very difficult for the person to develop pleasurable sexual response as an adult, except through years of therapy and a positive healing program.

An Act-centered theology generally does not center on pleasure, but rather on the procreational purpose of sexual intercourse. There is often little value placed on either the sexual response cycle or certainly the erotic stimulus pathway. A Relationship-centered theology has within it a potential for teaching and valuing pleasure as being central to the nature of sexual relationships. The problem, too often, is that the Relationship-centered person or religious group is shy about moving into the area of sexuality, thus contributing to the massive amount of ignorance, secretiveness and sexual dysfunction found in our culture. Unfortunately, I see too many persons from a Relationship-centered theology, who in the arena of their sexuality, live out an Act-centered sexual belief system. People are in moral conflict over their sexuality and the consequences are rampant ignorance and sexual dysfunction.

THE BELIEF SYSTEM OF THE THERAPIST

The belief system of the therapist is very important to the therapeutic process in working with issues of sexual health. The therapist, whether religious or not, is often caught in the same conflict between an Act-centered versus Relationship-centered value system. In approaching this matter, several key things should be kept in mind. First of all, it is important for the therapist to know his/her own value system. A therapist who is bound by an Act-centered value system will be limited to working only with clients who believe in an Act-centered theology. A Relationship-centered therapist can work with people all along the theological spectrum, but it is important for that therapist to understand and appreciate the client who comes from an Act-centered system. It is likely that a person with a sexual problem will feel more comfortable talking with a therapist from a Relationship-centered value system.

Second, it is important for the therapist to be as knowledgeable as possible regarding sexual issues. Education in sexuality is an important factor for the therapist working with sexual issues. The therapist who wishes to be helpful to clients who want to build meaningful relationships requires knowledge regarding all the issues mentioned in this chapter.

Third, it is important for the therapist to be able to help guide the client through the conflicts of his/her sexuality and his/her belief system. This needs to be done with a non-judgmental approach and understanding of the factors involved in the conflicts. It is most helpful when the therapist comes from a Relationship-centered faith stance. The therapist with the Act-centered theology usually feels bound to the Act-centered stance on the various issues.

Fourth, to keep up-to-date with all the new knowledge in the field of human sexuality, it is important for the therapist to relate to at least one of the professional human sexuality organizations.[1]

Finally, the therapist needs to educate the client about sex and the various religious value systems. Most people are sexually ignorant and do not understand the differences in religious value systems. They are in need of correct information if they are to achieve sexual and spiritual health. Sex therapy is most helpful when it includes good sex education as well as knowledgeable information about various faith stances.

University of Pennsylvania Graduate School of Education
Philadelphia, Pennsylvania, U.S.A.

NOTE

[1] The important organizations are as follows: SIECUS (Sex Information and Education Council of the U.S.), 130 West 42nd St., Suite 2500, New York, N.Y. 10036. (212) 819–9770. SIECUS has an outstanding sexuality library and publishes a bimonthly newspaper, *SIECUS REPORT,* with the most current information about sexuality issues.

AASECT (American Association of Sex Educators, Counselors and Therapists), 435 No. Michigan Ave., Suite 1717, Chicago, Ill. 60611–4067. (312) 644–0828. AASECT is the organization that certifies sex educators, counselors, therapists and supervisors. It publishes a monthly newspaper, *Contemporary Sexuality,* and a quarterly journal, *Journal of Sex Education and Therapy.*

SSSS (Society for the Scientific Study of Sex), P.O. Box 208, Mt. Vernon, Iowa 52314. (319) 895–8407. SSSS publishes an important academic journal, *Journal of Sex Research* and sponsors regional and annual meetings presenting current research in the field of sexology.

BIBLIOGRAPHY

1. Calderone, M.S. (ed.): 1974, *Sexuality and Human Values,* Association Press, New York.
2. Fisher, H.: 1990, Lecture on 'The Evolution of Monogamy, Adultery, and Divorce', given at the national meeting of the *Society for the Scientific Study of Sex,* Minneapolis.
3. Francoeur, R.T.: 1991, *Becoming a Sexual Person,* 2nd ed., MacMillan Publishing Co.,

New York.
4. Holland, J.M. (ed.): 1981, *Religion and Sexuality,* The Association of Sexologists, San Francisco.
5. Kaplan, H.S.: 1979, *Disorders of Sexual Desire,* Simon and Schuster, New York.
6. Masters, W. and Johnson, V.: 1966, *Human Sexual Response,* Little, Brown & Co., Boston.
7. Masters, W. and Johnson, V.: 1970, *Human Sexual Inadequacy,* Little, Brown & Co., Boston.
8. Money, J. and Ehrhardt, A.: 1972, *Man & Woman Boy & Girl: The Differentiation and Dimorphism of Gender Identity from Conception to Maturity,* Johns Hopkins University Press, Baltimore.
9. Money, J. and Tucker, P.: 1975, *Sexual Signatures: On Being a Man or a Woman,* Little, Brown & Co., Boston.
10. Stayton, W.R.: 1980, 'A Theory of Sexual Orientation: The Universe as a Turn On', in W.R. Stayton (ed.), *Topics in Clinical Nursing* 1 (4), 1–7.
11. Stayton, W.R.: 1985, 'Religion and Adolescent Sexuality', in R.B. Shearin (ed.), *Seminars in Adolescent Medicine* 1 (2), 131–137.
12. Stayton, W.R.: 1985, 'Alternative Lifestyles: Marital Options', in D.C. Goldberg (ed.), *Contemporary Marriage,* Dorsey Press, Homewood, Ill., pp. 241–260.
13. Stayton, W.R.: 1989, 'A Theology of Sexual Pleasure', *American Baptist Quarterly* 8 (2), 94–108.

MARGRETTA DWYER

TO WHERE HAVE WE COME: NOISY PHILOSOPHICAL PONDERINGS OF A QUIET MIND. AN EPIPHANY

Epiphany = A revelation, a kind of sudden brightness that lights up the landscape *of a mind*, of a community, of a whole social order.

INTRODUCTION

"Yes, I have been sexual with my 13-year-old daughter, but I never had vaginal intercourse with her – just anal intercourse for the years 11–13."

"Why? – Why?"

"I follow the rules of my religion; it is very important that my children marry as virgins. Our church requires it."

*****

"... my sexual relations with my wife have been poor – little or no sex, so I turned to my daughters."

"Why? – Why? Why not another woman?"

"My religion forbids extra-martial sexual activity. Premarital sex and extramarital relationships are wrong."

*****

Answers from a sex offender population. Religion turned into religiosity; mythical motifs carefully pursued so as to relate to the crucial personal cycle of dying, resurrecting and returning to heaven. Reason closed down, and rules blindly obeyed even though they no longer feel right. How can this be?

Society's codes, life, and institutions require myths/stories that offer real life models to us, but psychologically to be effective the models must be appropriate to the time and culture in which we live [2]. Past virtues, such as crusades and forced baptisms, may be the vices of today, while past vices, such as lending money, become today's necessities.

> The moral order (must) catch up with the moral necessities of today; ... the old-time religion belongs to another age, another people, another set of human values, another universe. By going back you ... are out of sync with history ([2], p. 13).

To live out of sync with present history, with present/personal experiences is to divide one's self, to live two lives. Possibly nowhere does this happen more than in the area of sexuality. Today, sexuality suffers severely from rigid rules

Ronald M. Green (ed.), Religion and Sexual Health, 219–223.

imposed on itself. Rules which were born and reside in technological, social and cultural conditions which no longer exist [8]. The context in which these rules were born is probably not questionable. Old regulations served their purpose in a time when it was believed that every sexual act had could have tremendous life and death implications. These were the times when the brightest were the medical personnel, promulgating rules regarding the body and its fluids. With no other place to turn for bodily/sexual information, church personnel used the medical myths, to create sexual rules for faithful Christians. Fortunately, the medical field developed new information about the body, but unfortunately, the Church held to the old information. Sexual regulations continued in a tradition that could not allow itself to pay attention to new experiences in new circumstances. Fear ensnared many to continue venerating the old; but as ecclesiastical creditability lessened, honesty led others to uncover a new church in the space of their hearts. When an established church loses credibility in one major area, the people quicken their pace, heeding their hearts, and manifest what appears to be "massive apostasy (with) ... notable decline in religious devotion and belief" [7]. When the people are able mentally to fracture legalism in any one major area such as sexuality (an example from Catholicism would be contraception – by 1980 approximately 75% of Catholic women of childbearing age used birth control; other religions, too, have their Achilles heel which led their faithful to fracture legalism) they are able spiritually to pursue the experiences of their heart in an ethical manner. This takes precedence in their lives. A journey is begun – for them – for posterity; for their heirs or family to follow.

TRADITIONAL RULES' UNDERPINNINGS

We need to explore the mythological underpinnings of sexual prohibitions in order to trace the moral pathways by which we have arrived at our sexual regulations. We must also understand the fear which keeps one enduring the dominant rules that make sexual behavior unenjoyable or unacceptable.

Understanding the *his*tory/*her*story on which we base our present day decisions is necessary for understanding how over-vigilant superegos develop at a very young age [11]. What possibly could have led to naming our sexual fulfillment as a "remedy for concupiscence"? [4]. (Concupiscence: lechery, depravity; crave, longing or wish for – these are terms drawn from around the world.) And so we live with our normal desires being called *depravity, concupiscence.*

Medical antisexualism antedates the 1500s. It was believed that too much loss of semen could weaken the man and even produce a form of castration. Strength would be lost; even today this prevails in sports medicine where competitors debate about abstaining from sex before competition for fear of weakening the body [10]. It followed naturally that masturbation must be forbidden if it depletes the life sources.

Further, it was believed in the 1700s that each sperm contained a "little person" and because this was so, to spill sperm unnecessarily would be destroying "people." (Only as recently as the 1920s were hormones extracted in chemical form, leading to the recognition that semen was not made from neurine fluid, robbing nerve tissue and draining brain fluids from the body [10].) With such frightening information, appropriate condemnation must be forthcoming. The church needed to condemn masturbation. Her credibility depended on it. Wet dreams were condemned as undisciplined lust spilling over into sleep [10]. Conservation of semen was a sign of moral purity, and religion must appear pure.

So, immersed in the sexual myths, each subtly given to us in our youth, we step into the outside world and further learn that knowledge and myths are promulgated by THE POWERS. The Powers decide what will be promulgated and how it will be promoted. We quickly learned the powers were the medical profession and the clerics, for long periods the two most educated bodies in society.

In the 1500s the standard medical understanding of the body was based on "a plumber's view." Good health came by a balance between the sexual production and the discharge of fluids "in the pipes." One medical writer states that "too much frequency of 'embraces' dulls the sight to gaze, dulls the memory, induces gout, palsies, and effeminates the whole body and shortens life" (source unknown). Medical personnel thought in terms of the body, including sexuality, as a closed-energy system. Individuals ignoring this closed-energy system would be shortening their life if they engaged in *inappropriate sexuality or sexual experiences.*

To shorten one's life is akin to suicide; in this context it is understandable why religions must speak out.

Various positions for intercourse were forbidden and considered medically unsound because the semen needed to run with gravity. To have semen flowing counter to gravity could hinder conception. And thus, was born "the missionary position only" for sexual relations.

At intercourse, a woman must be "rested," "sober," and free from all mental worries or a weak child would be produced. Little sex should occur during the summer months since sex overheats the blood and would dry up the body. If sex occurred during menstruation, a diseased child would be produced. All these myths contributed to feelings of guilt and fear surrounding sexuality. Intercourse was forbidden during the period of breast feeding because it contaminated another fluid – the mother's milk, risking injury to the child.

By the 1800s a double standard emerged. If a husband was adulterous "forgiveness on the part of the wife is meritorious, while similar forgiveness on the part of the husband is degrading and dishonorable" (source unknown).

In 1873, E. H. Clarke [3] argued that if a woman during her childbearing years directed too much energy to any other organ – especially the brain – physical collapse and even death could occur. (Women were taught needlepoint, repetitive tasks, so as to keep the *blood fluid* from the brain.) Now women were beginning to enter higher institutions of learning, and a controversy resulted. Clarke's book declared higher education for women a health hazard. Co-education was unacceptable since it might interfere with the normalization of a girl's cycle if she associated every day with boys. (Enter: all girls schools!)

A 1901 medical textbook [6] continued to advocate padlocking the foreskin of the male or the inner labial genital lips of the female by punching holes into these areas to prevent masturbation. These procedures were considered more humane than the former chastity belts or the suturing nearly closed of the vagina and the anus. A 1920 medical text reiterated the teaching of the 1901 volume [9].

In 1954, Gordon Rattray Taylor stated that when society restricts sexuality more than the human constitution can stand, one of three things occurs: "Either people will defy the taboos, or they will turn to perverted forms of sex, or they will develop psychoneurotic symptoms" ([12], p. 54).

TO WHERE HAVE WE COME?

Time has proven Taylor both right and wrong. He is right that these three situations will come to exist; he is wrong (if he meant to imply so) that only one will come about at a time.

Although burdened by guilt and shame, some people have still defied the taboos surrounding sexuality and have been alienated from family, friends, jobs, and normal living situations in order to follow their beliefs or orientation (e.g., homosexuals); others have turned to perverted forms of sexuality – risking arrest, loss of reputation and other painful experiences by repressing normal sexual desires and acting out abnormally (exposing, voyeuring, incest, etc.); still others have developed psychoneurotic sexual symptoms seen today in many sexuality clinics throughout the country – extreme sexual guilt, suppression of normal sexual desires, and problems of sexual dysfunction within marriages. Other symptoms show themselves by inappropriate boundary issues in the work place (sexual harassment), or sexual compulsivity issues in the bookstores and adult theaters, sometimes leading to the need for long-term treatment.

Pleasure and power are strangely at odds in the area of sexuality. Although power is all around us and comes from everywhere, the powers that be in the sexual arena have been medical and religious. "Power is not an institution, and not a structure; neither is it a certain strength we are endowed with; it is the name that one attributes to a complex strategical situation in a particular society" ([5], p.93). Who conceived the power? One must wonder if the underlying theme of power repressing sexuality is not of our own punitive doing, a result of the refusal to see and understand the new circumstances that are being brought to light for us all. What would it mean for each of us "to articulate an ethic that frees, nurtures, and sustains sexual energy for the sake of personal wholeness?" [1].

Given our background, it is difficult to recognize our sexual energies as something to be cherished and enjoyed. For most people I have met, the struggle is intense as well as guilt ridden.

But to continue to be disconnected and guilty about an integral part of ourselves is to live out of sync with our present history and ourselves. It is to perpetuate psychoneurotic symptoms and perverted forms of sex. To say yes to our deepest cravings, without guilt, can expedite peace within ourselves and contribute to eliminating sexually offending behavior in society.

Sexual offending behaviors remain a mystery even to those working in the field for some years, but the issues of guilt, religion turned into religiosity, and rules blindly obeyed cannot be ignored when dealing with issues of paraphilias. Sexual offending behavior has many components, not all of which we scientifically understand at this time. Sexual energy – a primal energy – extends deep into each of us. Keeping this energy caged seems to influence sexual offending behavior or paraphilias, rather than aid "normal" turn-ons. Working with sexual offenders leads one to know that a key element of offending behavior is sexual repression. Repressing a primal energy such as sexual energy, appears, from all we can tell, to be dangerous, and yet we seem to fear this primal energy called "sexual." We therefore carry on wars within ourselves to harness these powers unnaturally and ignore the revelations, the "Epiphanies" that can light up our minds.

University of Minnesota Medical School
Minneapolis, Minnesota, U.S.A.

BIBLIOGRAPHY

1. Andolsen, B.H.: 1992, 'Whose Sexuality? Whose Tradition? Women, Experience, and Roman Catholic Sexual Ethics', (in present volume, 55–77).
2. Campbell, J.: 1988, *The Power of Myth,* Doubleday, New York.
3. Clarke, E.H.: 1873, 'Sex in Education: Or a Fair Chance for the Girls', (no other publication data found).
4. Curran, C.: 1992, 'Sexual Ethics in the Roman Catholic Tradition', (in present volume, 17–35).
5. Foucault, M.: 1980, *The History of Sexuality,* Vol. 1, Vintage Books, New York.
6. Gould-Pyle: 1901, (title and publisher unknown).
7. Greeley, A.M. et al.: 1976, *Catholic Schools in a Declining Church,* Sheed and Ward, Kansas City, Missouri.
8. Green, R.M.: 1987, 'The Irrelevance of Theology for Sexual Ethics', in E.E. Shelp (ed.), *Sexuality and Medicine,* Vol. II, D. Reidel Publishing Co., Dordrecht, pp. 249–270.
9. Jefferis: 1920, *Search Light on Health,* (publisher unknown).
10. Money, J.: 1985, *The Destroying Angel,* Prometheus Books, Buffalo, NY.
11. Simpson, W.S. and Ramberg, J.A.: 1992, 'The Influence of Religion on Sexuality: Implications for Sex Therapy', (in present volume, 155–165).
12. Taylor, G.R.: 1954, *Sex in History,* Thames and Hudson, London.

NOTES ON CONTRIBUTORS

Barbara Hilkert Andolsen, Ph.D., is Helen Bennett McMurray Visiting Professor of Social Ethics at Monmouth College, West Long Branch, New Jersey.

Vern L. Bullough, Ph.D., is SUNY Distinguished Professor of History, SUNY College, Buffalo, New York.

Charles E. Curran, S.T. D., is Scurlock Professor of Ethics, Southern Methodist University, Dallas, Texas.

Margretta Dwyer, M.A., is Coordinator of the Sex Offender Treatment Program, University of Minnesota Medical School, Minneapolis, Minnesota.

George Edwards, Ph.D., is Professor Emeritus of New Testament, Louisville Presbyterian Theological Seminary, Louisville, Kentucky.

Ronald M. Green, Ph.D., is John Phillips Professor of Religion, Dartmouth College, Hanover, New Hampshire.

Christine E. Gudorf, Ph.D., is Professor of Theology, Xavier University, Cincinnati, Ohio.

Harold I. Lief, M.D., is Professor Emeritus of Psychiatry at the University of Pennsylvania and Psychiatrist Emeritus to the Pennsylvania Hospital.

James B. Nelson, Ph.D., is Professor of Christian Ethics, United Theological Seminary of the Twin Cities, New Brighton, Minnesota.

Joanne A. Ramberg, Ph.D., is Director, Mental Health Program, Washburn University, Topeka, Kansas.

David E. Richards, B.D., a Bishop of the Episcopal Church, is currently associated with the firm of *Performance* which provides pastoral counseling services directed especially to clergy and their families.

William S. Simpson, M.D., is Director, Center for Sexual Health, The Menninger Clinic, Topeka, Kansas.

Julian W. Slowinski, Psy.D., is in private practice in clinical psychology and is Senior Clinical Psychologist at the Pennsylvania Hospital.

William Stackhouse, Ph.D., is a counseling psychologist in New York City.

William R. Stayton, Th.D., is in private practice as a sex counselor in Wayne, Pennsylvania. He serves on the faculty of the Human Sexuality Program at the University of Pennsylvania Graduate School of Education, Philadelphia, Pennsylvania.

Ronald M. Green (ed.), Religion and Sexual Health, 225.

INDEX